D1131177

A Practical Guide to

INSTRUMENTAL ANALYSIS

Ernő Pungor

CRC Press

Boca Raton   Ann Arbor   London   Tokyo

Library of Congress Cataloging-in-Publication Data

Pungor, E. (Ernő)
  A practical guide to instrumental analysis / E. Pungor
     p.   cm.
  Includes bibliographical references and index.
  ISBN 0-8493-8681-0
  1. Instrumental analysis. I. Title.
  [DNLM: 1. Hepatitis B virus. QW 710 G289h]
  QD79.I5P86 1994
  543—dc20

                                        94-28608
                                          CIP

No claim to original U.S. Government works
International Standard Book Number 0-8493-8681-0
Library of Congress Card Number 94-28608
Printed in the United States of America   2 3 4 5 6 7 8 9 0
Printed on acid-free paper

Preface

The field of instrumental analysis is connected with many areas of science. Consequently, the collection of the necessary knowledge and the understanding of the fundamentals of the methods require a great deal of work from the beginner. The aim of this book is to facilitate this work by providing a short introduction to each technique as well as to introduce the reader to the practice of the methods. The selection of the fields of instrumental analysis and of the experiments included is the result of long academic teaching practice and experimentation. I think that the experiments will enable the starting analyst to gain the knowledge and skill necessary to apply the various techniques.

It may be worthwhile to say some words about the birth of the book. I have been teaching various chapters of instrumental analysis since 1949, and prepared the first laboratory manual in 1953 for my students. Then, together with my colleagues we have changed, extended, and modified the contents to suit practical needs. This work has been continued till the present day, which explains the long list of my colleagues, old and new, who contributed to the book, to whom I wish to express my thanks: J. Balla, L. Bezur, K. Erőss Kiss, Zs. Fehér, O. Gyimesi, E. Gráf Harsányi, M. Hangos-Mahr, T. Kántor, G. Nagy, R. Paulik, L. Pólos, K. Tóth, and first of all Jenő Fekete for his contribution in preparing the chapter of chromatography.

I owe special thanks to A. Hrabéczy-Páll for her help in editing this book.

I hope the book will be of assistance to all those who wish to get acquainted with the fields of instrumental analysis that are included. (Naturally, there are other important fields of instrumental analysis, such as surface analytical and structure analysis techniques, for which the reader is referred to other books.)

I wish the reader success in using the book.

Ernő Pungor
Professor & Head
Analytical Department
Technical University
Budapest

Table of Contents

Chapter 23. Automatic Laboratory Analyzers

Part I

Electroanalytical Techniques

Chapter 1

Potentiometry

Principle of the Technique

Techniques based on the measurement of potential of a sensor are termed potentiometry. In a potentiometric type of sensor a membrane or sensor surface (indicator electrode) acts as a half-cell, generating a potential proportional to the logarithm of the analyte activity (concentration). The indicator electrode is connected directly (or through a salt bridge) with a reference electrode to form a galvanic cell. The electromotive force (EMF) or cell voltage of the galvanic cell thus obtained is measured in such a manner that no current flows through the cell. For a complete electrochemical cell the cell voltage is

$$E_{cell} = E_{ind} - E_{ref} + E_j$$

where E_{cell} is the cell voltage or the EMF, E_{ind} is the potential of the indicator electrode, E_{ref} is the potential of the reference electrode, and E_j is the liquid-junction potential.

The potential of the indicator electrode is given by the following equation:

$$E_{ind} = Constant + 2.303RT/zF \log a$$

where $2.303RT/zF$ = Nernst factor (its value at 25°C and in the case of monovalent ions is 59.16 mV), z = charge of the ion measured, and a = activity of the ion measured.

Potentiometric methods belong to two major types:

1. Direct potentiometry. Direct measurements of concentrations or activities. In this case the ion activity or ion concentration is determined by means of a calibration curve or the standard addition technique.

2. Indirect potentiometry or potentiometric titration. The component to be determined is titrated with a suitable titrant and the indicator electrode is used to follow the changes in potential in the course of the titration.

1.1. Determination of pH in Fruit Juice

Principle of the Determination

The term pH is simply a mathematical symbol of convenience, widely accepted and firmly established for expressing the acidity of aqueous solutions. It is the negative logarithm of activity of the hydrogen ion:

$$pH = pa_{H^+} = -\log a_{H^+}$$

where $a_H{}^+ = f_+[H^+]$ (f_+ = single ion activity coefficient).

However, single ion activity coefficients cannot be measured directly; only the mean activity coefficients are available ($f\pm$).

Since the pH values defined by this formula cannot be determined exactly, it was necessary to introduce the practical ("operational") pH scale, which is based on standards. The most commonly used pH scale is the NBS (National Bureau of Standards) pH scale.

For the determination of the practical pH values, the following equation can be used:

$$pH(x) = pH(s) + (Ex - Es)/2.302RTF^{-1}$$

where pH(x) is the pH of the test solution and pH(s) is the pH of the standard.

Ex and Es are the EMF values measured in the test solution and in the standard, respectively. It has been suggested that in practical pH determinations with a glass electrode the best approach is to use two standards from the NBS scale, one below and one above the pH value of the test solution. The EMF of the cell containing test solution x (E_x):

$$\text{glass electrode} \mid \text{solution} \mid \text{reference electrode}$$

is measured and the EMF values E_1 and E_2 are similarly measured using the same cell containing standard solutions S_1 and S_2, respectively, having pH values pH_1 and pH_2 on either side as near as possible to the pH of x (=pHx).

$$(pH_x - pH_1)/(pH_2 - pH_1) = (E_x - E_1)/(E_2 - E_1)$$

Apparatus

mV- pH meter
Glass electrode
Reference electrode (Ag/AgCl or calomel)

Chemicals

pH buffers

Sample

Fruit juice

Procedure

Prepare the pH buffer solutions and measure the Emf values in the standards and the test solutions. Repeat the measurements on the pH scale after adjusting the pH meter to the pH value corresponding to the pH of one of the standards (e.g., pH = 7), then to the pH value of a second standard (e.g., pH = 2). Finally, the pH of the sample is measured.

Evaluation

Calculate the pH on the basis of the millivolt measurements and compare it with that obtained by direct pH measurement.

1.2. Determination of Fluoride in Toothpaste with Fluoride Ion-Selective Electrode Using Standard Addition

Principle of the Determination _____

The solid-state fluoride electrode has found extensive use in the determination of fluoride in a variety of materials. A total ionic strength adjustment buffer (TISAB) is used to adjust all unknowns and standards to essentially the same ionic strength; therefore, the concentration of fluoride can be measured and evaluated. The pH of the buffer is about 5, a level at which F^- is the predominant fluorine species. The buffer also containes complexing agent, which forms stable chelates with iron(III) and aluminum(III), thus eliminating their possible interference.

Solutions

1. TISAB. Buffer solution for 15 to 20 determinations can be prepared by mixing (with continuous stirring) 57 ml of glacial acetic acid, 58 g of NaCl, 4 g of cyclohexylamine dinitrilo-triacetic acid, and 500 ml distilled water in 1-l beaker. Cool the content and add 6 M NaOH until a pH of 5.0 to 5.5 is reached. Dilute 1 l with water and store in a plastic bottle.
2. Standard fluoride solution (2×10^{-1} mol/l). Dry a quantity of NaF at 110°C for 2 h. Cool in a desiccator, then weigh the appropriate amount of salt into a 1-l volumetric flask. (Caution! NaF is highly toxic.) Dissolve in water, dilute to the mark, mix well, and store in a plastic bottle.

Apparatus

> pH meter
> Solid-state fluoride electrode
> Reference electrode (Ag/AgCl double junction or calomel
>   electrode)
> Magnetic stirrer

Procedure

1. Determination of the slope of the fluoride calibration curve. Prepare 100, 100-ml calibration solutions in the concentration range of 10^{-5} to 10^{-2} mol/l fluoride containing fifty 50-ml TISAB buffer. From the calibration curve the (E vs. $-\log c_{F^-}$) the slope is determined.

$$S = \Delta E / \Delta(-\log c)$$

2. Determination of fluoride in toothpaste. Weigh about 1 g of toothpaste into a 250-ml beaker. Add 50 ml of TISAB solution and boil for 2 min with good mixing. Cool, then transfer the suspension quantitatively to a 100-ml volumetric flask, dilute to the mark with distilled water, and mix well. Measure the E_1 (mV) value in a given volume (V) of this unknown solution, then add to this solution a known volume (v) of a fluoride standard of c_{add} concentration. Measure again the E_2 (mV) in the cell.

Evaluation

From the measured data calculate the concentration of fluoride in the sample using the following equations:

$$E_1 = E^\circ - S \log c_x$$

$$E_2 = E^\circ - S \log\left[(c_x \times V + c_{add} \times v)/V + v\right]$$

From the two equations c_x and the fluoride concentration of the toothpaste sample are calculated. Report the parts per million of F^- in the sample.

1.3. Determination of the Stoichiometric Dissociation Constant of Acetic Acid by Potentiometric Titration

Principle of the Determination

The dissociation constants of weak acids and bases can be determined on the basis of their potentiometric titration curves. The dissociation equilibrium of a monovalent weak acid can be described by the following equation:

$$HA \rightleftharpoons H^+ + A^-$$

The dissociation constant is

$$K_d = \frac{[H^+][A^-]}{[HA]}$$

from which

$$\log K_d = \log[H^+] + \log\frac{[A^-]}{[HA]} = \log\frac{[A^-]}{[HA]} - pH$$

$$\log\frac{[A^-]}{[HA]} = \log K_d + pH \tag{1}$$

Accordingly, if the portion of the titration curve up to the equivalence point is transformed into a $\log\frac{[A^-]}{[HA]}$ vs. pH plot, a straight line of a slope of +1 is obtained whose intersection with the abscissa gives $-\log K_d$ (Figure 1.1).

Let us introduce the degree of titration, defined as

$$x = \frac{\text{amount of titrant in the solution}}{\text{amount of titrand originally present}} = \frac{V(cm^3)}{V_e(cm^3)}$$

where V_e is the volume of titrant required up to the equivalence point and V is any volume of titrant added to the titrand before the equivalence point.

At the beginning of the titration $V = 0$, that is, $x = 0$; at the equivalence point $V = V_e$ and $x = 1$.

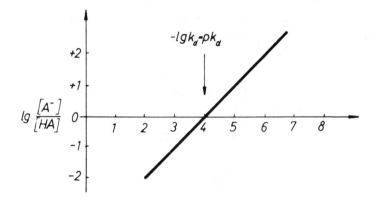

FIGURE 1.1 $\log \frac{[A^-]}{[HA]}$ vs. pH diagram.

Using the degree of titration (x) and initial acid concentration (C_{acid}), the left-hand side of Equation 1 can be transformed:

$$\log \frac{[A^-]}{[HA]} = \log \frac{x \times C_{acid} + [H^+] - [OH^-]}{(1 - x)C_{acid} - [H^+] + [OH^-]} \qquad (2)$$

[H^+] and [OH^-] can be calculated from the pH value belonging to a point of titration at a degree of titration x. Equation 2 can be simplified if $C_{acid} \geqq 10^{-2}\,M$, since in the 4 to 10 pH range the quantities [H^+] and [OH^-] can be neglected. Thus,

$$\log \frac{[A^-]}{[HA]} = \log \frac{x}{1 - x} \qquad (3)$$

Comparing Equation 3 with Equation 1 we have

$$\log \frac{x}{1 - x} = \log K_d + pH \qquad (4)$$

In the half-titrated solution x = 0.5; thus according to Equation 4

$$\log 1 = \log K_d + pH_{0.5}$$

$$pH_{0.5} = -\log K_d = pK_d$$

Accordingly, the pH in the half-titrated solution is equal to the dissociation exponent of the weak acid.

Apparatus

Precision pH meter
Combination glass electrode
Burette, 12 cm^3
Pipette, 10 cm^3
Beaker, 200 cm^3
Magnetic stirrer

Chemicals

0.1 *M* acetic acid
0.1 *M* carbonate-free sodium hydroxide standard solution
Buffer solutions in the pH range of 4 to 6.

Procedure

Standardize the pH meter using buffer solutions of known pH. In this manner, a titration curve of adequate accuracy can be obtained using the given electrode. Transfer a 10.00-cm^3 aliquot of 0.1 *M* acetic acid into a 200-cm^3 beaker using a pipette; dilute it to about 100 cm^3 with distilled water. Immerse the combination glass electrode (Figure 1.2) and put the stirrer into the solution. Read the pH of this solution. Titrate with 0.1 *M* sodium hydroxide standard solution, adding the titrant in 0.5-cm^3 portions. After each addition, wait for the pointer to come to a standstill and read the pH. Reduce the portions to 0.1 to 0.2 cm^3 toward the end-point of the titration.

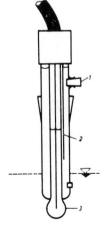

FIGURE 1.2
Combination glass electrode. (1) Opening for filling the reference electrode, (2) reference electrode, and (3) glass bulb (indicator electrode).

Evaluation

Plot the pH vs. the volume of solution added. Determine V_e, the volume belonging to the equivalence point, and plot the linearized titration curve corresponding to Equation 4. Read $pK_d = -\log K_d$. Calculate the concentration of the acetic acid (mg/100 cm^3) in the original solution.

1.4. Determination of Potassium in Blood Serum by Direct Potentiometry with an Ion-Selective Electrode (ISE)

Principle of the ISE Based Method

The potassium ISE is a liquid-type membrane electrode. The ISE which incorporates a neutral carrier-based ionophore, valinomycin, or a bis-crown-ether derivative, separates the solution being measured from the electrode internal filling solution. The membrane components — ionophore, plasticizer, matrix material, membrane anionic mobile sites (lipophilic salts, such as NaTPB) — are insoluble in water. The potentiometric response of the ion-selective membrane electrode is due to the selective interaction of the ionophore and the potassium ion present in the solution. The boundary potentials thus exist at both the external and the internal interfaces. The inner boundary potential is kept constant by fixing the concentration of potassium in the internal filling solution (0.001 M KCl). The outer boundary potential and consequently the membrane potential then varies depending on the potassium activity in the test or sample solution. The potential across the membrane, i.e., the membrane potential, is measured by a millivolt meter of a high input impedance (10^{-12} Ω) in contact with a reference electrode on either side of the membrane, usually a SCE or Ag/AgCl reference electrode.

The membrane potential response to potassium ion activity, a_{K+}, in the test or sample solution according to the Nernst equation is

$$E = E_{const.} + 0.0591 \log a_{K+}$$

The potassium ion activity is related to the potassium concentration by the ionic activity coefficient, γ_K; $a_{K+} = \gamma_{K+}C_{K+}$

The activity coefficients depend greatly on the total ionic strength of the sample solution and can be approximated by the Debye-Hückel equation as

$$\log \gamma_{\pm} = \frac{A|z_+ z_-|\sqrt{I}}{1 + Ba\sqrt{I}} + cI$$

$$I = 0.5 \sum_n c_n z_n^2$$

where A is −0.509 at 25°C, in water; B is 0.328; z_+, z_- is charge numbers of cation and anion of the electrolyte; I is ionic strength of electrolyte; a, C is constants (for potassium a = 3.65; C = 0.015); c_n is concentration of any ion in the solution; and z_n is charge of any ion in the solution.

The potassium electrode also responds to cations that are complexed by the ionophore and thus can compete in the generation of the boundary potential. These cations can cause an interference if present in sufficiently large concentration as compared to that of the principle ion. The response of an ISE in the presence of interfering ions can be described by the Nicolsky equation:

$$E = E_{const.} + 0.0591 \log\left[a_{K^+} + \sum_n K_{K,J}\, a_J^{1/z_j z_j}\right]$$

where a_J is interfering ion, J, with charge z_j; $K_{K,J}$ is the potentiometric selectivity coefficient of the electrode for potassium over J; and n is number of interfering ions.

The selectivity coefficient is a measure of the extent of the interference caused by interfering ion, such as J.

Apparatus

pH meter with expanded scale (or specific ion meter)
Potassium ISE (home prepared)
Reference electrode; double junction SCE or Ag/AgCl, KCl
   with 1.0 M LiOAc
Salt bridge electrolyte
Magnetic stirrer, magnetic bars
Beakers (50 ml)
Volumetric flasks (100 ml)
Pipettes (10 ml)

Chemicals

0.1 M KCl standard solution
Ionic strength adjusting electrolyte (blood serumlike electrolyte
   [BSE], 140.0 mM NaCl, 24.0 mM NaHCO₃, 0.6 mM MgCl₂,
   1.1 mM CaCl₂)
1.0 M LiOAc solution
0.001 M KCl solution
Sample to be analyzed: blood serum

Reagent for Membrane Preparation

> Polyvinyl chloride, PVC (Chromatographic grade, e.g., Fluka 81392)
> Dioctyl sebacate, DOS (Selectrophore®, e.g., Fluka 84818)
> Potassium tetrakis(4-chlorophenyl)borate, KTpClPB (additive) (Selectrophore®, e.g., Fluka 60591)
> BME-44 (Selectrophore®, e.g., Fluka 60397) or
>   valinomycin (Selectrophore®, e.g., Fluka 60403)
> Tetrahydrofuran, THF (e.g., Fluka 87369)

Procedures

Membrane Preparation

2.0 mg ionophore, 65.0 mg PVC, 120.0 mg plasticizer, and 70 mol% (relative to the ionophore) additive were dissolved in 2.0 ml THF and poured into a glass ring (diameter 28 mm). A flexible membrane of approximately 0.2-mm thickness remains after the evaporation of the THF.

Electrodes

Solvent polymeric membranes (diameter 7.0 mm) are incorporated into Philips IS-561 electrode bodies or mounted on a Tygon tube of 7 mm outer diameter which has been fixed at the end of a glass tube (approximately 10 cm long). The membrane can be fixed by dipping the end of the Tygon tubing into tetrahydrofuran to soften, attaching the membrane to it, and allowing it to dry. As an inner reference electrolyte, use $10^{-3}\,M$ KCl; as an inner reference electrode, an Ag-AgCl wire can be employed.

Potential Response of the Electrodes

1. Prepare 100.0 ml of each of the following KCl solutions by serial dilution of the 0.1 M KCl standard solution: 10^{-2} to $10^{-6}\,M$. Measure the potential for each stirred solution 2 min after immersion. Plot the measured potential values (mV) vs. log a_K^+ (calculated with the help of the Debye-Hückel equation). Calculate the slope of the plot and compare it with the theoretical value. Determine the range over which the response is linear to the log a_K^+; estimate the detection limit.
2. Prepare 100.0 ml of each of the following KCl solutions by serial dilution of the 0.1 M KCl standard solution with BSE: 10^{-2} to $10^{-6}\,M$ KCl/BSE. Measure the potential for each stirred solution 2 min after immersion. Plot the measured potential values (mV) vs. log a_K^+. Calculate the slope of the plot and determine the range over which the response is linear to the log a_K^+; estimate the detection limit. Compare all the parameters with the ones obtained with the use of pure KCl solutions.

Determination of the Selectivity Factor $K_{K,Na}$

Prepare 100.0 ml of 0.1 M KCl solution and 100.0 ml of 0.1 M NaCl solution with distilled water. Measure the cell potential values in both solutions separately allowing time to obtain a stable reading. Calculate the selectivity factor as

$$\log K_{K,Na} = \frac{E_{Na} - E_K}{0.0591}$$

Calculate the error of the potassium determination by considering the determined selectivity data and 5×10^{-3} M KCl and 0.14 M NaCl as physiological average concentration values:

$$K_{K,Na} = \frac{a_K}{a_{Na}} \cdot \frac{p}{100} \%$$

where p is the error of the determination.

Calibration Curve Method

Transfer exactly 25.0 ml serum sample into a 50-ml plastic beaker and measure the cell potential. Determine the concentration of potassium from the calibration curve. Do not discard: the solution can be used for further measurements.

Standard Addition Method

Add 250 µl 5×10^{-1} M KCl to 25.0 ml serum sample; mix and read the new potential value. Repeat the standard addition step at least five times. Record the new potential values after each addition. Calculate the concentration in the serum sample with the help of the following equation:

$$c_x = \frac{c_s V_s}{\left(V_s + V_x\right) \times 10^{-\Delta E / 0.0591} - V_x}$$

where c_x = unknown concentration, c_s = concentration of the standard, V_x = volume of unknown sample, V_s = volume of standard solution, ΔE = (E after standard addition) − (E of unknown).

Compare the potassium concentration of the sample determined by the two methods.

1.5. Redox Titration in the Presence of a Platinum-Calomel Electrode Pair

1.5.1. Titration of Iron(II) with Cerium(IV) Standard Solution

Principle of the Determination _____

In strongly acidic medium at room temperature, the titration is carried out according to the following reaction:

$$Fe^{2+} + Ce^{4+} = Fe^{3+} + Ce^{3+}$$

Apparatus

> mV-pH meter
> Bright platinum electrode
> Calomel reference electrode
> Magnetic stirrer
> Automatic burette (10 ml)

Chemicals

> 0.005 M cerium(IV)sulfate standard solution
> 1.00 M sulfuric acid

Procedure

Bring a 5- to 10.00-cm^3 aliquot of the iron(II) mol/l solution to a 200-cm^3 beaker by means of a pipette; acidify it with 30 cm^3 of 1.00 M sulfuric acid solution. Fill up the solution to about 120 cm^3 with distilled water and immerse the platinum and reference electrodes. Connect the electrodes to the mV-pH meter. Titrate with 0.05 M cerium(IV)sulfate solution added in 0.25-cm^3 portions at the beginning and in 0.1-cm^3 portions toward the end-point of the titration.

Evaluation

Plot the potentiometric titration curve and calculate the iron(II) concentration of the sample solution. The evaluation can be carried out by determining the inflection point of the titration curve (Figure 1.3) or plotting the derivative potentiometric titration curve (Figure 1.4) and determining the maximum value of this curve.

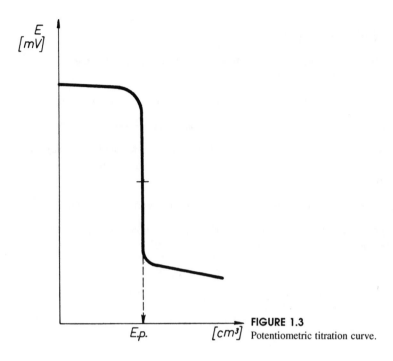

FIGURE 1.3
Potentiometric titration curve.

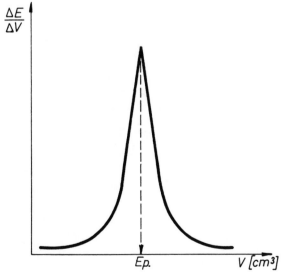

FIGURE 1.4
Derivated potentiometric
titration curve.

1.6. References

Ammann, D., Morf, W. E., Anker, P., Meier, P. C., Pretsch, E., and Simon, W., *Ion-Selective Electrode Rev.,* 5, 3, 1983.

Bailey, P. L., *Analysis with Ion-Selective Electrodes,* Heyden and Son, London, 1976.

Cammann, K., *Working with Ion-Selective Electrodes,* Springer-Verlag, Berlin, 1979.

Meier, P. C., Ammann, D., Morf, W. E., and Simon, W., in *Medical and Biological Applications of Electrochemical Devices,* Koryta J., Ed., John Wiley & Sons, New York, 1980.

Oehme, F., *Ionenselektive Elektroden,* Dr. Alfred Hüthig Verlag, Heidelberg, 1986.

Weissberger, A. and Rossiter B. W., Eds., *Physical Methods of Chemistry,* Part IIA, Wiley Interscience, New York, 1971.

Chapter 2

Biamperometric Titration (Dead-Stop End-Point Detection)

Principle of the Method _____

The dead-stop method is based on the phenomenon of polarization. A constant potential difference is applied to two identical electrodes immersed into the solution. If the solution contains a reversible redox couple, a current will flow through the cell, which can be measured by a microamperemeter. The solution is titrated under stirring and the current is read after each addition of the titrant. The schematic diagram of the circuit used is shown in Figure 2.1.

If the redox system present is irreversible and the sum of the overpotentials of the anodic and cathodic processes exceeds the external voltage applied to the electrodes, in principle no current can pass through the cell; in practice the current is very small.

2.1. Determination of Iodine by Titration with Sodium Thiosulfate Solution, Using Biamperometric End-Point Detection

Principle of the Determination _____

When titrating iodine with the thiosulfate and applying a voltage of 20 mV to the electrodes, the sum of the overvoltages of the reaction $I_2 + 2e = 2I^-$ at the cathode and of the reaction $2I^- = I_2 + 2e$ at the anode is 15 mV. On titrating with

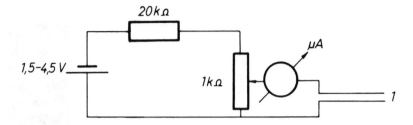

FIGURE 2.1
Basic circuit diagram of the dead-stop end-point detection method, 1-point electrodes.

an irreversible redox system of 2-V overvoltage: $S_4O_6^{2-}/S_2O_3^{2-}$ the current decreases up to the equivalence point, then, as iodine disappears, drops to nearly zero at the equivalence point.

Apparatus

2-V lead storage battery
Potentiometer unit with connecting wires and switch
Galvanometer
Mechanical stirrer
mV meter
Titration vessel with platinum sheet electrodes
Microburette
Pipettes, 1 and 2 cm^3

Chemicals

0.01 N potassium hydrogen iodate
0.01 N sodium thiosulfate
2 N hydrochloric acid
Solid potassium iodide

Procedure

Bring 1.00 or 2.00 cm^3 of the 0.01 N KH $(IO_3)_2$ solution into a titration vessel with a pipette, add about 0.5 g of solid KI, 10 cm^3 of water, and 1 cm^3 of 2 N HCl. Apply a voltage of 20 mV to the electrodes and titrate the iodine formed with 0.01 N Na$_2$S$_2$O$_3$ solution, adding the standard solution in 0.1-cm^3 portions under stirring from a microburette. Read the galvanometer after each addition. Continue the titration by adding a few 0.5-cm^3 portions of the titrant after the current has reached a constant level.

Evaluation

Plot the titration curve: galvanometer reading vs. volume of titrant. Calculate the potassium hydrogen iodate content of the original solution (mg/cm^3).

2.2. Determination of Water in Methyl Alcohol by the Karl Fischer Method

Principle of the Determination _____

The Karl Fischer method can be used for the accurate and rapid determination of small amounts of water (up to 200 mg). The basis of the method is that sulfur dioxide reacts with iodine in the presence of water in a reaction leading to equilibrium:

$$SO_2 + I_2 + 2H_2O \rightleftharpoons SO_4^{2-} + 2I^- + 4H^+$$

The equilibrium can be shifted to the right-hand side, to the formation of strong acid by using pyridine. The reaction is completely unequivocal for water; methyl alcohol is used as a solvent for the sulfur dioxide, iodine, and pyridine.

The Karl Fischer standard solution, which contains about 1 mol of iodine, 3 mol of sulfur dioxide, 10 mol of pyridine, and 50 mol of methanol, can be used for the direct titration of water in any solution which does not react with sulfur dioxide and/or iodine. Solutions containing aldehydes or ketones cannot be titrated, since they bind sulfur dioxide.

As shown by the composition, the water equivalent of the standard solution is determined by the iodine content. For a fresh solution it is 4 to 5 mg of water/cm^3.

The water equivalent of the solution decreases on storing, due to absorption of water and to light-induced side reactions. Therefore, it is advisable to standardize the solution daily.

With colorless solutions the excess of iodine can be observed visually, the color changing from yellow to brown after the equivalence point. With both colorless and colored solutions, the biamperometric (dead-stop) and bipotentiometric end-point detection can be applied successfully, using two platinum indicator electrodes. In addition to high sensitivity, the advantage of the method is that it provides an electrical signal which enables the titration to be automated.

A Labor Aquameter (automatic titrimeter) is used in which a constant current of 100 µA from a generator is forced through the cell (bipotentiometry). The valve of the automatic burette is controlled by the potential difference between the electrodes. The valve closes when the titration of water has been completed and the burette can be read (Figure 2.2).

Apparatus

> Labor Aquameter
> Vessel with dropper for weighing water
> Analytical balance
> Pipettes, 1, 2, and 5 cm^3
> Graduated cylinder, 25 cm^3

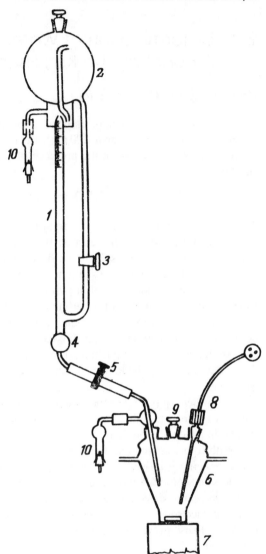

FIGURE 2.2
Automatic bipotentiometric
titrator (Labor Aquameter).
(1) burette, (2) reservoir,
(3) stopcock, (4) magnetic valve,
(5) tube clamp, (6) titration
vessel, (7) magnetic stirrer,
(8) electrodes, (9) ground glass
stopper, and (10) tubes filled
with silica gel.

Chemicals

Karl Fischer solution
Methyl alcohol
Methyl alcohol sample to be titrated

Procedure

Assemble the Aquameter according to Figure 2.2, start the apparatus according
to the operating manual, and fill up the burette and reservoir with Karl Fisher

solution. Pour 10 cm³ of methyl alcohol (solvent) into the titration vessel and start the titration. Adjust the dropping rate to rather fast using a tube clamp and adjust the end-point potential on the basis of the titration result (see operating manual).

Add 1 drop of distilled water from the weighing bottle to the just titrated, i.e., water-free, solution and determine the weight of the drop (20 to 30 mg). Care should be taken that the drop falls directly into the solution and not onto the wall of the vessel. Fill the burette with the reagent up to the "O" overflow pipe and titrate the water immediately.

If some water gets on the ground neck, on the wall of the vessel, on the electrodes, or on the side tube for titrant introduction, wash it into the solution prior to titration with a few cubic centimeters of methyl alcohol using a pipette.

Titrate to accurately water-free. Do three to five parallel titrations adding further known amounts of water.

Calculate the

$$\frac{\text{mg } H_2O}{\text{cm}^3 \text{ Karl Fischer solution}} \quad \text{water equivalent values}$$

Calculate the average value from single values which differ by less than 0.1 mg/cm³.

In order to determine the water content in the sample, prepare a water-free solution from about 10 cm³ of methanol in the titration vessel by titration with Karl Fischer reagent. Then add 1 cm³ of sample using a pipette and make an informatory titration. Then titrate 1-, 2-, or 5-cm³ aliquots according to the water content so that the standard solution requirement falls between 5 and 15 cm³. Make three to five parallel determinations, adding the new sample portions to the solution already titrated.

Evaluation

Calculate the water content of the methyl alcohol sample from the average of the results of parallel titrations and from the mg H_2O/cm³ water equivalent of the standard solution, in wt % H_2O and ppm.

2.3. References

Stock, J. T., *Amperometric Titrations,* Interscience, New York, 1965.

Chapter 3

Voltammetry, Polarography

Voltammetry is an electroanalytical measuring technique which can be used for the quantitative determination of reducible or oxidizable components. The technique is based on the study of current vs. potential relationships. In voltammetry the working electrode is polarizable, i.e., by changing its potential according to a special time program, in a certain range the potential of the electrode will change according the applied potential program. The reference electrodes used in voltammetry are nonpolarizable, i.e., their potential remains constant in the course of the voltage change in the measuring cell. (For ensuring the controlled working electrode potential generally a third, so-called auxiliary electrode is also used.) In the course of the potential change components present in the solution to be studied can be oxidized or reduced; as a result of these processes current is flowing through the cell. The current vs. potential relationships are called voltammetric curves.

In voltammetry different working electrodes and different potential vs. time programs are used (see Figure 3.1). The most commonly known and used technique is polarography, where the working electrode is a dropping mercury electrode (DME), and the potential vs. time program is linear.

The polarographic techniques can be used mainly for the determination of heavy metals and organic molecules on the basis of their reduction on the dropping mercury electrode. The reduction can occur in the molecule itself (e.g., molecules with double bonds) or in the reducible group of the molecule (e.g., nitro, nitroso, or carbonyl group).

If diffusion conditions are ensured the so-called classical polarography is suitable for the determination of analytes in the concentration range of 5×10^{-3} to 5×10^{-5} mol/l. In addition to the dropping mercury electrode, other different types of mercury electrodes are used. For the determination of oxidizable

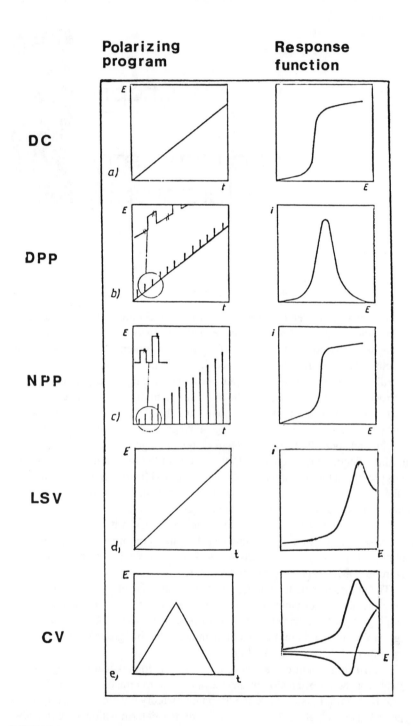

FIGURE 3.1   The polarizing programs and response functions used in voltammetry.

components generally noble metal and carbon (graphite) working electrodes are available. Electrodes of a second kind or metal electrodes of high surface area can be used as reference electrodes, while a platinum electrode of high surface area is most commonly used as an auxiliary electrode.

Since the polarographic current intensity (the limiting current) is a well-defined function of the concentration of electroactive components (see the Ilkovic equation), the concentration of sample solution can be determined either by calibration or by standard addition technique. In addition to these, polarography can be used for the following titrations, too (e.g., amperometric titrations).

3.1. Preparation of Solutions for Polarographic Analysis

3.1.1. Experiment to Study Polarographic Maxima

Principle of the Measurement _____

The aim of the experiment is to study polarographic maxima and the way of maximum suppression.

Primary maxima appear at the beginning of polarographic waves; they are usually high in dilute, low-conductivity solutions. They are attributed to convection in the vicinity of the surface of the mercury drop working electrode. The maxima — with the exception of a few special cases — interfere in quantitative determinations; hence, it is advisable to suppress them. Maxima can be eliminated by the addition of polarographically inactive surfactants. Gelatin solution is the most widely used maximum suppressant.

In the experiment potassium chloride is used — in a concentration lower than usual — as a supporting electrolyte, and the maximum appearing on the polarographic wave of dissolved oxygen is studied.

Apparatus

Polarograph

Chemicals

10 mol/l potassium chloride solution
0.5% gelatin solution
Distilled mercury

Procedure

Pour mercury into the measuring cell (e.g., a 25-ml beaker, very carefully, over a tray) to cover the bottom, then add 10 to 15 ml 10 mol/l potassium chloride solution and place the cell on the polarographic stand. Put on the cell cover and introduce the dropping mercury working electrode and the wire sealed into a glass tube ending in a small platinum peak to ensure connection of the mercury pool reference electrode to the polarograph. Take care to allow sufficient distance between the mercury pool and the dropping mercury electrode so that drop formation is not hindered. Position the connecting wire in a way that the platinum peak does not touch the bottom of the cell. Obviously, the platinum must not contact the solution. An electrode of the second kind can also be used as reference electrode. Adjust the appropriate drop-time (3 to 4 s/drop) by adjusting the height of mercury reservoir. Set the appropriate sensitivity, potential range (0.0 to –2V), potential scan (polarization rate, e.g., 250 mV/min), damping, and record the polarographic curve. The maximum is clearly seen on the curve. Add four or five drops of gelatin solution (0.5%) to the solution in the cell, mix, and record the polarographic curve under the same conditions as before. The maximum suppressing effect of gelatin is clearly seen.

3.1.2. Role of Oxygen in Polarography

Principle _____

In this experiment the polarographic reduction waves of oxygen is studied along with the ways of eliminating the interference by oxygen.

Aqueous solutions in contact with air contain a significant amount of dissolved oxygen which is polarographically active, i.e., it is reduced at the dropping mercury electrode. The polarogram of oxygen consists of two well-separated diffusion-controlled waves. The first wave is due to the reduction of oxygen to hydrogen peroxide. The reaction proceeds in acid solution as follows:

$$O_2 + 2H^+ + 2e^- = H_2O_2$$

The second wave is the result of the reduction of hydrogen peroxide to water which proceeds in acid solution as follows:

$$H_2O_2 + 2e^- + 2H^+ = 2H_2O$$

Accordingly, the oxygen concentration of solutions can be measured polarographically. However, oxygen interferes in the polarographic determination of various materials. In practice, with the exception of a few special cases, oxygen has to be removed before polarographic determinations. Oxygen can be removed by bubbling an inert gas (nitrogen, argon, helium) through the solution or by chemical reaction, using a reductant. Even traces of oxygen have an adverse effect on the detection limit of polarographic determinations.

In the majority of cases the solution is purged with a stream of purified nitrogen, taking care to avoid any solvent loss and consequent concentration change. In alkaline solution sodium sulfite may be used for removing oxygen.

Apparatus

As in Experiment 1.

Chemicals

As in Experiment 1.

Procedure

This experiment is the continuation of the previous one. The polarogram obtained is evaluated by determining the half-wave potentials of the two waves. Then nitrogen is passed through the solution for 8 to 10 min and the polarogram is recorded under the same conditions as before. The polarogram consists of two parts, the first being the residual current and the second the steeply rising part indicating the reduction current of the potassium chloride background electrolyte.

Attention! In polarography, if the aim is not the measurement of oxygen, care should be taken to remove oxygen continuously. During polarographic measurement the inert gas is passed over the solution, before and between measurements through the solution.

3.2. Polarographic Spectrum of a Solution Containing Thallium(I), Nickel(II), and Manganese(II) Ions

Principle

The aim of the experiment is to study the polarographic behavior of polarographically active materials when present together, to estimate the resolution of the method. Polarography does not belong to the group of high-resolution quantitative analytical methods. Thus, its use in qualitative analysis is limited. However, in a number of cases simultaneous determinations are possible. An example of this is the following experiment.

Apparatus

Polarograph
10 ml pipette
200 µl pipette

Chemicals

> Supporting electrolyte: 1:1 (v/v) mixture of 0.5 mol/l ammonium
> hydroxide and 0.5 mol/l ammonium chloride solution
> 0.01 mol/l thallium(I) nitrate solution
> 0.01 mol/l nickel(II) sulfate solution
> 0.01 mol/l manganese(II)sulfate solution
> 0.5% (w/v) gelatin solution
> Distilled mercury

Procedure

The polarographic cell is prepared as in the previous experiment; 10.00 ml of the supporting electrolyte is introduced, four or five drops of gelatin solution are added, and the solution is purged with nitrogen for 8 to 10 min. Introduce the electrodes to the solution. The polarogram is recorded in the potential range of (0 to −2.0) V at a scan rate of about 0.25V/min, at a suitable sensitivity and damping. The polarogram should have the form expected for the supporting electrolyte.

Then 200 µl of 0.01 mol/l thallium(I) nitrate solution is added and after removal of oxygen by purging with nitrogen for 1 min the polarogram is recorded under the same conditions as before. The experiment is repeated after addition of 200 µl of 0.01 mol/l nickel(II) sulfate solution and again after the addition of 200 µl of manganese(II) sulfate solution.

The experiment is repeated four or five times after addition of 200-µl portions of 0.01 mol/l nickel(II) sulfate solution.

Evaluation

1. Prepare a calibration curve for nickel by plotting the current corresponding to the wave height against the nickel concentration of the solution. Observe the validity of the Ilkovic equation.
2. Determine the half-wave potentials of the three ions by graphical method. Calculate the half-wave potentials vs. saturated calomel electrode (SCE). (It is known that the half-wave potential of thallium(I) is −0.49 V vs. SCE, fairly constant in the presence of various background electrolytes. Use this knowledge in determining the half-wave potentials of the other two ions studied, if other than saturated calomel electrode is used as reference electrode.)

3.3. Separation of Polarographic Waves by Complexation

Principle

The aim of the experiment is to study the role of complexing agents in polarography in increasing the resolution of the technique.

The effect of complexing agents on the half-wave potentials of the reduction of metal ions is known. This phenomenon can be utilized in the resolution of overlapping waves. The experiment serves to demonstrate this effect.

Lead(II) and thallium(I) ions are reduced at the dropping mercury electrode practically at the same potential in acid and neutral solution; the polarographic waves of the two ions overlap. Under these conditions these ions cannot be determined by polarography. However, in alkaline solution the lead forms a hydroxo complex, while the thallium remains in free ionic form. Thus a difference of about 300 mV exists between the half-wave potentials which is large enough to allow two well-defined and separated waves to be obtained in a solution containing the two ions in about the same concentration.

Apparatus

> Polarograph
> 10.00-ml pipette
> 200-μl micropipette

Chemicals

> 0.1 mol/l potassium nitrate solution (supporting electrolyte)
> 0.1 mol/l thallium(I) nitrate solution
> 0.1 mol/l lead(II) nitrate solution
> 0.5% (w/v) gelatin solution
> Distilled mercury
> Sodium hydroxide (solid)

Procedure

Pipette 10.00 ml of the supporting electrolyte into the polarographic cell; add 200 μl of lead(II) nitrate solution and four or five drops of gelatin solution. Introduce to the cell the dropping mercury working electrode as well as the reference and auxiliary electrode.

Bubble nitrogen through the solution for 8 to 10 min and record the polarogram starting at –0.2 V, scan range 2.0 V, scan rate 0.25 V/min at an appropriate sensitivity and damping. Then add 200 μl of 0.1 mol/l thallium(I) nitrate, deoxygenate for 1 min, and record the polarographic curve under the same conditions as before. Add about 0.5 g solid sodium hydroxide to the solution and mix by bubbling nitrogen until complete dissolution. Record the polarogram again similarly as before.

Evaluation

Determine the half-wave potentials graphically and the shift in the values due to the formation of hydroxo complex.

It is interesting to carry out a logarithmic analysis of the two waves, to determine the number of electrons (n) involved in the reduction of thallium and the irreversibility factor (α) for lead.

3.4. Determination of the Drug Content of a Pharmaceutical Preparation

As it was mentioned earlier the polarographic technique can be used for the determination of organic molecules capable of being reduced or oxidized in the potential range allowed by the electrodes and supporting electrolytes applied. With various mercury electrodes in most cases the concentration of reducible molecules or ions can be determined.

In this experiment the drug content of a 1,4-benzodiazepine compound containing pharmaceutical preparation is determined. (It is supposed that the active ingredient is Nitrazepam and the tablet to be examined contains 10 mg of the drug.)

Apparatus

> Polarograph
> Polarographic cell with DME or SMDE stand
> Reference and auxiliary electrodes
> Nitrogen (or argon) gas cylinder
> Volumetric flasks of 50-ml volume
> 5.00-ml pipettes
> Mortar
> Funnel with filter paper
> Beakers (25 and 50 ml)

Chemicals

> Nitrazepam substance
> Nitrazepam containing tablet (drug content: 10 mg)
> Ethanol
> Supporting electrolyte which is 0.2 M in NaCl and 10^{-3} M in HNO$_3$

Procedure

Prepare a 10^{-2} mol/l stock solution from Nitrazepam drug (dissolve 0.1405 g material using ethanol as solvent applying a 50-ml volumetric flask). Prepare calibration solutions in the concentration range 10^{-3} to 10^{-4} mol/l by dilution of the stock solution with distilled water. Take the polarograms of the calibration solutions. Pulverize three Nitrazepam-containing tablets one after the other, using about 10 ml of ethanol, and transfer the suspension to a volumetric flask of 50 ml. Complete the volume to 50 ml with distilled water. Filter about 12 ml of the suspension; transfer 10 ml of the solution by a pipette to the polarographic cell. Take the polarogram. Repeat the procedure with the other two tablets.

Evaluation

Determine the limiting current graphically in the case of the calibration as well as the sample solutions. Draw the calibration graph on the basis of the current intensities obtained with the calibration solutions. Determine the drug content of the tablets on the basis of the calibration line.

3.5. Comparison of Different Polarographic Techniques

3.5.1. Study of a Polarogram of a Solution Containing Pb2+, In3+, Cd2+, and Zn2+ in Different Operation Modes

Principle

In addition to classical polarography other polarographic (voltammetric) techniques are used in the analytical practice to enhance the capabilities of classical polarography.

The detection limit of polarography is determined by the residual current, more precisely the condenser current. Utilizing the different time dependences of the condenser current and Faraday current different measuring techniques were developed to lower the detection limit of polarography.

The so-called sampled DC (its earlier name was Tast polarography) is based on the fact that the diffusion current at the dropping mercury electrode is increasing according to $t^{1/6}$, while the condenser current is decreasing according to $t^{-1/3}$ in the course of the lifetime of the mercury drop. Accordingly, the current is measured in a narrow interval of the last period of the lifetime of the mercury drop. In this time interval the ratio of the diffusion current and the condenser current is the most favorable.

The potential vs. time relationships and the corresponding current vs. electrode potential functions for the pulse techniques can be seen in Figure 3.1.

As it can be seen, at both techniques potential pulses are applied to the working electrode, once in the lifetime of the mercury drop. The duration of these pulses is about 50 to 100 ms.

In the case of normal pulse polarography (NPP) linearly increasing potential pulses are applied, and the current is measured at the end of the time interval of the pulse. In this time interval the value of the condenser current is small, since it decays in the course of the pulse exponentially. The shape of the pulse polarogram is similar to the classical one; the limiting current can be described as follows:

$$i_{NP} = nF\overline{A}C_o \sqrt{\frac{D}{\pi t}}$$

where n is the number of electrons taking part in the electrode process; $\overline{A}$ is the surface area of the electrode; D is the diffusion coefficient of the component taking part in the electrode process; and t is the time, elapsed from the beginning of the application of potential pulse.

As it can be seen, the limiting current of NPP is according to the Cottrell equation.

In differential pulse polarography (DPP) — as it can be seen in the figure — the polarizing potential is increased linearly (as in the case of classical direct current polarography, DC), and the potential pulses of about 10 to 100 mV are added to the increasing potential. The current intensity is measured in a narrow interval of time before the application of the pulse and at the last part of the pulse duration. The instrument shows the differences of these two measured current intensities as a function of the working electrode potential. As a consequence of this, the DPP curve is a peak type. The current intensity at the maximum can be described as follows:

$$i_{DP} = \frac{(nF)^2}{4RT} \overline{A}\tau_3 \ C_o \Delta E \sqrt{\frac{D_o}{\pi t}}$$

where ΔE is the amplitude of the potential pulse and τ_3 is the duration of the pulse.

It can be concluded that both pulse techniques are suitable for the determination of concentration of unknown concentrations.

It should be noted that the techniques discussed above can be carried out only if the dropping of mercury and the potential pulses are synchronized. This can be achieved in two ways: either the drop formation and detachment are controlled in accordance with the measurement program or the dropping is used for timing the potential pulses and current measurement. The use of controlled dropping mercury electrodes (static mercury drop electrode, SMDE) offers several technical advantages; hence, this electrode is the most widely applied working electrode in voltammetric pulse techniques.

The aim of the experiment is the comparison of the different polarographic techniques.

Apparatus

> Polarograph with SMDE stand
> Polarographic cell with reference as well as platinum auxiliary electrodes
> 10.00-ml pipettes
> Micropipette (with adjustable volume or volume of 200 μl)

Chemicals

> Potassium chloride, reagent grade
> Standard solution I, which is 10^{-2} mol/l in lead nitrate, 10^{-2} mol/l in indium nitrate, 10^{-2} mol/l in cadmium acetate, and 10^{-2} mol/l in zinc sulfate
> Standard solution II, which is 10^{-4} mol/l in lead nitrate, 10^{-4} mol/l in indium nitrate, 10^{-4} mol/l cadmium acetate, and 10^{-4} mol/l zinc sulfate 10^{-4} mol/l cadmium acetate solution

Procedure

Prepare the supporting electrolyte, which is 0.1 mol/l KCl solution. Pipette 10 ml of the supporting electrolyte into the polarographic cell. Remove the oxygen from the solution by bubbling N_2 or argon gas through the solution for 10 min. Record the current vs. potential relationship of the supporting electrolyte at different current sensitivities in the potential range of 0 to -1.4 V vs. saturated calomel reference electrode.

Add 200 μl of Pb^{2+}-; 200 μl of In^{3+}-; 200 μl of Cd^{2+}-; and 200 μl of Zn^{2+}-ion containing standard solution I (concentration: 10^{-2} mol/l) the supporting electrolyte, bubble the N_2 gas through the solution for about 1 min, and take the polarograms in the different operation modes (DC, DC_T, NPP, DPP). To obtain a well-defined polarogram change the different instrument parameters. After rinsing thoroughly the electrodes and N_2-introducing tube, pipette 10 ml supporting electrolyte into another polarographic cell.

Repeating the procedure written above eliminate the oxygen from the solution and take the polarogram of the supporting electrolyte, but at higher sensitivity, than that used in the previous experiment.

Add 200 μl of Pb^{2+}-; 200 μl of In^{3+}-; 200 μl of Cd^{2+}-; and 200 μl of Zn^{2+}-ion containing standard solution II (concentration: 10^{-4} mol/l) to the supporting electrolyte, bubble the N_2 gas through the solution for about 1 min, and record the polarograms in the different operation modes (DC, DC_T, NPP, DPP).

In the next experiment add another portion (200 μl) of 10^{-4} mol/l Cd^{2+}-containing solution and after N_2 bubbling again take the polarograms.

Evaluation

1. Compare the performance of the four polarographic techniques (determination of lowest concentrations, which can be determined; selectivity or resolution).
2. Determine the half-wave as well as the peak potentials of the components in the background electrolyte used.
3. Calculate the cadmium concentration of the solution containing 200 μl Cd^{2+} ion by the standard addition technique and compare the nominal and measured values of the concentration.

3.6. Determination of Chlorpromazine Content of a Pharmaceutical Preparation by Standard Addition Technique Applying a Graphite Working Electrode

Principle

A great number of the organic compounds cannot be reduced or oxidized using mercury working electrode, since the potential range allowed by the mercury electrode in different background electrolytes is not wide enough for the oxidation. The anodic potential range is available by the application of noble metal or carbon (graphite) electrodes. The platinum, gold, and other metal electrodes are used mainly in nonaqueous media, while the graphite electrodes are used in aqueous solutions.

Different graphite electrodes have been developed in the last 30 years; the aim was to obtain nonporous graphite electrodes. Nowadays the most frequently used electrodes are the carbon paste and the so-called Glassy Carbon ones. By the application of these electrodes different organic compounds can be determined, e.g., compounds containing amino, phenolic OH groups or aromatic molecules with heteroatoms.

Solid electrodes can be used as working electrodes in different voltammetric techniques; in the following determination the direct current voltammetry will be used. Since the potential program of the technique is the same as in the case of DC polarography, the name of this technique is linear sweep voltammetry (LSV). The shape of the voltammetric curve obtained is the peak type (see Figure 3.1). The first two parts (residual current, linearly increasing part of the curve) corresponds to the same processes as those of polarography, while the third, decreasing part can be attributed to the depletion of the electroactive material in the vicinity of the electrode during the voltammetric run. The reason for this process is the constant surface area of the electrode and the linear diffusion of the electroactive component (which is a fairly good assumption). For the peak current (I_p in A) Randles and Sevcik derived the following equation:

$$I_p = k n^{3/2} D^{1/2} A v^{1/2} c$$

where k is a constant, called the Randles Sevcik constant (its value is 2.7×10^2); n is the number of electrons taking part in the electrode process; A is the electrode area in cm^2; v is the rate of potential change (scan rate, rate of polarization) in V/s; and c is the concentration of the electroactive component in mol/l.

Application of electrodes of constant surface area involves the advantage that the value of the residual current is one order of magnitude lower than that on a dropping mercury electrode. This is an important feature of these kinds of electrodes because this way the detection limit is smaller by about of one order of magnitude. The disadvantage of the use of electrodes of constant surface area is that in certain cases it can be contaminated because the product of the electrode process used for analytical purposes stays on the surface of the electrode forming an insoluble layer. The renewal of the surface can be carried out by polishing the electrode; however, there may be some difference in the renewed and the original surface area (5 to 10%). The so-called filming effect decreases with the decrease of the concentration of the analyte.

The aim of the experiment to gain some experience in the use of solid voltammetric electrodes. The model compound is chlorpromazine, a phenothiazine derivative:

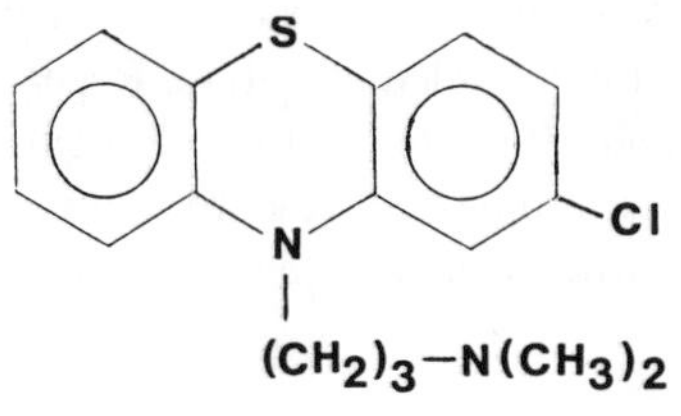

A free radical is forming from the molecule in the first step of the reaction, but the reaction can go further, and the end product is a sulfoxide.

Apparatus

         Polarograph
         Magnetic stirrer
         Stirring bar

Mortar
Funnel with filter paper
Measuring cell (50-ml beaker)
Graphite working electrode
Saturated calomel reference electrode
Platinum auxiliary electrode
10.00-ml pipette
200-µl micropipette

Chemicals

Supporting electrolyte (0.01 mol/l in hydrochloric acid and
0.1 mol/l in potassium chloride)
Chlorpromazine substance
Chlorpromazine containing pharmaceutical preparation (drug
content: 25 mg)

Procedure

Transfer 10.00 ml of the supporting electrolyte to the voltammetric cell. Adjust the appropriate sensitivity and record the voltammogram starting from a voltage of +0.2 V in the scan range 2 V in a stationary solution. It is advisable to adjust a higher scan rate than with polarography (e.g., 50 m V/s) and no damping. Thus, the voltammogram of the supporting electrolyte is obtained.

Pulverize the chlorpromazine-containing pharmaceutical preparation and dissolve it in the supporting electrolyte. Pour the suspension into a 100-ml volumetric flask and complete it to 100 ml. Filter a few milliliters (e.g., 10) of the suspension and pipette 1 ml of the filtered solution into the supporting electrolyte. Mix for 30 s and wait for 10 s until convection completely ceases. Then the voltammogram is recorded as before. Repeat the recording after 30 s mixing the solution and 10 s waiting time. Record three voltammograms. After addition of 1 ml chlorpromazine standard solution (concentration: 25 mg/ 100 ml) the solution is stirred again and the voltammogram is recorded four or five times.

Evaluation

Determine the peak current of the voltammogram before and after the standard addition. Calculate the concentration of the sample solution and then the real drug content of the pharmaceutical preparation.

It is interesting to study the effect of polarization rate; for this purpose record the voltammogram of the chlorpromazine-containing solution applying different scan rates. Plot the dependence of the peak current intensity on the square root of the scan rate.

Another interesting experiment to study the effect of the depletion of the component to be studied is in the vicinity of the electrode during the voltammetric run. For the observation of this phenomenon record three voltammograms in the chlorpromazine-containing solution, but without mixing between the subsequent scans.

3.7. Cyclic Voltammetry

Principle _____

If a stationary electrode is used in a stationary solution, the product of the reaction which serves as the basis of voltammetric determination stays in the vicinity of the electrode surface for a certain time. This allows various experiments to be carried out, one of which is the so-called cyclic voltammetry. Using this technique, conclusions can be drawn concerning the nature of the electrode process. The shape of the polarizing potential vs. time relationship is an isoscale triangle in the case of the cyclic voltammetry, i.e., a cyclic voltammogram is prepared by starting the potential scanning in one direction, then continuing it in the opposite direction.

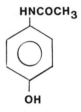

The aim of the experiment is to provide a quick insight into the practice of cyclic voltammetry. One of the components to be studied is chlorpromazine, while the other is the N-acetyl-p-aminophenol:

Apparatus

As in Section 3.5.

Chemicals

Chlorpromazine solution, 0.01 mol/l
N-acetyl-p-aminophenol solution, 0.01 mol/l

Procedure

The voltammetric cell is prepared as in Section 3.5, using the supporting electrolyte and 400 µl chlorpromazine stock solution. The recording is started under the same conditions (in the anodic direction), but the direction of voltage scanning is changed at a suitable point of time (soon after the decomposition of the supporting electrolyte starts).

Repeat the measurement using 10.00 ml of the supporting electrolyte and 400 µl of N-acetyl-p-aminophenol (Paracetamol) stock solution.

Comparing the two curves it can be concluded that the product of chlorpromazine (free radical) formed in the first step of the oxidation can be reduced, while the product of N-acetyl-p-aminophenol cannot be reduced because of chemical complication including the oxidation process.

Evaluation

Determine the potentials belonging to the cathodic and anodic peaks and the difference (peak separation, which is 59.16/n for reversible electrode processes: n is the number of electrons involved in the reaction).

3.8. References

Bond, A. M., *Modern Polarographic Methods in Analytical Chemistry,* Marcel Dekker, New York, 1980.

Sawyer, D. T. and Roberts, J. I., Jr., *Experimental Electrochemistry for Chemists,* John Wiley & Sons, New York, 1974.

Vydra, F., Stulík, K., and Juláková, E., *Electrochemical Stripping Analysis,* Ellis Horwood, Chichester, England, 1976.

Chapter 4

Coulometry

Principle of the Technique

The analytical application of Faraday's law was first proposed by Szebellédy and Somogyi. Coulometry is the general term designating an electroanalytical method which consists in the measurement of the quantity of electricity used either to transform the substance to be determined in a redox reaction or to produce a reactant of an analytical reaction. Thus the species to be determined reacts either at one of the electrodes or with a reagent produced by electrolysis. The second method is also referred to as coulometric titration.

This method is based on Faraday's second law:

$$m = \frac{MQ}{nF}$$

where m is the amount of substance transformed in the electrolysis, M is the molecular weight of the substance, n is the number of electrons involved in the electrode reaction, Q is the amount of electricity used in the reaction in coulombs, F is the Faraday number (96.500 coulombs = 26.8 A hours).

A basic requirement in coulometric analysis consists in selecting an electrode reaction which takes place with 100% current efficiency as the amount of electricity is employed to calculate the quantity of the substance to be determined.

Current efficiency can be affected by interfering reactions, e.g., the solvent or the supporting electrolyte can react at the electrode, the electrode material can react with some components of the solution, or a secondary reaction can take place involving the product of electrolysis.

A large number of analytical reagents can be prepared by electrolysis and employed in coulometric analysis. Coulometry also permits the production of various reagents which, because of their instability, cannot be prepared by conventional means, e.g., Cu(I), Ag(II), etc. Coulometry is advantageous in micro or ultramicro analysis as very small amounts of electricity can be measured with precision. The thickness of various metal oxide or sulfite layers can also be determined by the coulometric method.

Coulometric titrations can be performed in two ways: the reagent is produced either in the solution containing the substance to be determined or it can be prepared in a separate cell and continuously introduced into the solution.

Coulometric analysis can be performed

1. At a constant current (amperostatic or galvanostatic coulometry). In this case the measurement of the amount of electricity consists of the measurement of time. The equivalence point of the titration can be indicated by various methods such as photometric, potentiometric, or amperometric techniques.
2. At a controlled potential (potentiostatic coulometry). In this case the potential of the working electrode is kept constant while the current decreases because of the decrease in the concentration of the reacting species. The amount of electricity is obtained by integrating the current vs. time function (Figure 4.1). In this technique, end-point detection is substituted by observing the decrease in the intensity. However, the residual current must be taken into account.

Both techniques can readily be automated.

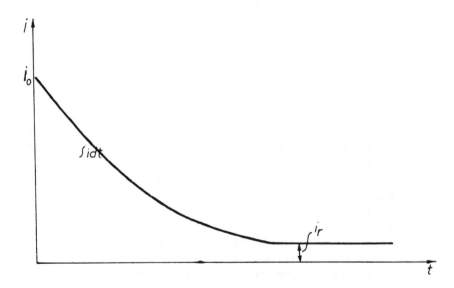

FIGURE 4.1
Current vs. time functions (i_o: initial current, t: time, i_r: residual current).

4.1. Coulometric Determination of Chloride Ions

Principle of the Determination _____

Silver ions are analytically generated from metallic silver, at a constant current. Silver ions react with the chloride ion content of the solution, forming an AgCl precipitate. The end point is indicated potentiometrically.

Apparatus

Coulometric analyzer containing constant current generator, timer, and indicator circuits*
Cell with silver and platinum generator electrodes, Ag/AgCl indicator electrode, and a suitable reference electrode
1 cm^3 graduated pipette or micropipette
Magnetic stirrer

Solutions

0.01 *M* potassium chloride solution
1 *M* perchloric acid in 50% methyl alcohol as the supporting electrolyte

Procedure

The coulometric apparatus is set in operation according to the instruction manual. The cell is filled with 20 to 50 cm^3 supporting electrolyte and 0.1 cm^3 0.01 *M* KCl solution is added. The cell is assembled. The stirrer is set in operation and the constant current generator and the timer are simultaneously switched on. The potential of the indicator electrode is plotted as a function of time. When an automatic coulometric titrator is used the plot is employed to present the end-point value for subsequent automatic determinations.

4.2. Coulometric Determination of Thiosulfate

Principle of the Determination _____

Thiosulfate is titrated with iodine generated coulometrically from a KI solution at a constant current. End-point detection is performed either by biamperometric or bipotentiometric techniques.

---

* Recommended: Radelkis, Type OH-404.

Apparatus

Coulometric analyzer
Cell with two Pt foil electrodes (for reagent generation) and two
Pt wire electrodes (for end-point detection)

Solutions

Phosphate buffer (pH = 7.0)
1 *M* potassium iodide

Procedure

The coulometric apparatus is set in operation according to the instruction
manual. The cell is filled with 75 cm³ phosphate buffer and 25 cm³ 1 *M* KI
solution. The cell is assembled. The current generator and the timer are
simultaneously switched on.

The current or the voltage of the indicator electrode is plotted as a function
of time.

4.3. Coulometric Determination of Cu(II)

Principle of the Determination _____

Cu(II) ions are reduced to copper on a mercury cathode. Copper amalgam is
formed during this reaction. The determination is carried out by potentiostatic
coulometry; thus, there is no need for end-point detection.

Apparatus

Coulometric analyzer
Cell with Pt wire and Pt foil electrodes, calomel reference
electrode

Solution

1 *M* NH_4Cl + NH_4OH buffer solution

Procedure

The cell is filled with 100 cm³ of the buffer solution. Mercury is layered at the
bottom of the cell. The cell is assembled. The potentiostat is switched on; the
stirrer is started. The current is plotted as a function of time. The area under the
curve is determined.

4.4. Determination of Hydrochloric Acid by Controlled-Current Coulometry (Szebellédy's Method)

Theory

As it is well known, a coulometric titration involves the electrochemical generation of the titrant. The titrant reacts with the sample, and the end point is determined by an appropriate method. The result is calculated from the electric charge passed through the reagent-generating electrode using the Faraday equation.

To obtain accurate results we need to perform the reagent generation with 100% current efficiency. When an acid is titrated, the following electrode reactions are used for reagent generation:

$$H_2O + e^- \longrightarrow 1/2\,H_2 + OH^-$$

$$H_3O^+ + e^- \longrightarrow 1/2\,H_2 + H_2O$$

The reagent generating cathodic reaction can be easily carried out in stirred solution at a platinum electrode with high surface area. However, no reducible material should be present in considerable concentration since it would decrease the current efficiency.

When electrolysis is carried out, both cathodic and anodic processes are taking place simultaneously. In coulometry the exclusion of disturbing interactions of these is important. If a species, e.g., the product of an anodic oxidation, gets to the cathode area and undergoes an electrolytic reduction, this will decrease the current efficiency of the cathodic reagent-producing process. Hence, the two half cells often need to be separated. Current bridges with high electric conductivity are used for this purpose. If we use an indifferent anode, e.g., a platinum electrode, as a counter electrode in titrating acids with coulometric reagent generation, then on the anode the electrolysis will produce acid (or decrease basicity) in the following reactions:

$$3H_2O - 2e^- \longrightarrow 1/2\,O_2 + 2H_3O^+$$

$$2OH^- - 2e^- \longrightarrow 1/2\,O_2 + H_2O$$

The high mobility of the hydroxonium and hydroxide ions makes the effective separation of the two half cells in this case very difficult.

It was László Szebellédy who, in the early times of coulometric research, introduced dissolving silver anode and bromide analyte in the practice of coulometric acid determinations. In this case the product of the anodic process is silver ion which is precipitated by the bromide ions and thus cannot interact and disturb the titrant-generating cathodic processes:

$$Ag^0 - e^- \longrightarrow Ag^+$$

$$Ag^+ + Br^- \longrightarrow AgBr$$

Szebellédy titrated acids with this trick without separating the anodic and cathodic half cells. In our method we titrate hydrochloric acid samples, but we employ a cell with separated cathodic and anodic compartments.

For end-point indication, we employ potentiometric detection and methyl orange color indicator.

Apparatus

     Coulometric instrument, precision current generator, current integrator, or stop watch
     Built-in or separate pH meter for end-point detection
     Coulometric cell
     Two-compartment cell with separating sintered glass filter
     Magnetic stirrer
     Platinum tape reagent-generating electrode
     Silver rod electrode
     Combination glass electrode

Chemicals

     Potassium nitrate solution 0.5 mol/l
     Potassium bromide solution 0.5 mol/l
     Hydrochloric acid sample solution
     Hydrochloric acid standard solution 0.01 mol/l
     Methyl orange indicator solution

Procedure

Put together the coulometric cell according to Figure 4.2. Place the stirring bar into the cathode compartment. Introduce the potassium nitrate solution, a few drops of the indicator solution, and the sample to be titrated. The platinum reagent-generating electrode is in this compartment. The volume of the electrolyte depends on the cell size. The solution must cover the surface of the reagent-generating electrode and the current bridge. It is advantageous if we purge the carbon dioxide with nitrogen stream before and during measurements. Transfer a 1:1 mixture of potassium nitrate and potassium bromide solution by pipette to the anode compartment. The volume of this

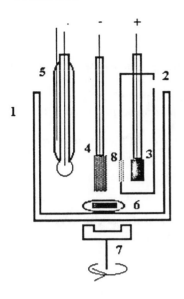

1-Coulometric cell (cathodic half)
2-Anodic half cell
3-Silver electrode
4-Platinum electrode
5-Combination glass electrode
6-Magnetic stirring bar
7-Magnet
8-Sintered glass filter

FIGURE 4.2
Coulometric cell.

mixture is determined by the size of the compartment. It should cover the functioning part of the electrode and should be in hydrostatic equilibrium with the catholyte. Place the silver electrode into this solution. During titrations an insulating silver bromide layer is building slowly around the silver electrode. It should be checked and removed from time to time. Place the glass electrode into the cathode compartment. Connect the electrodes to the current generator and to the pH meter. Determine the titration end-point pH. After this, introduce the sample. While employing continuous stirring adjust the current (0.5 to 30 mA/cm^2 electrode surface) and pass known amounts of electric charge through the cell. Measure the potential difference after each reagent-generating step between the end-point indicating electrodes and plot it against the sum of the electric charge passed through the cell. Approaching the end point the reagent-generating charge portions should be reduced. In this way the titration curve is prepared, the end point is determined, and from the Faraday equation the sample concentration is determined. The end-point potential is determined with a titration performed in the described way. The color indicator shows on which side of the titration is the mixture. In the case of an apparatus with automatic end-point controller we determine the end point manually, adjust it, and then with serial sample addition we can determine several samples with convenient push-button control.

To avoid extensive time and silver electrode consumption, or too fast overtitration, a previous calculation is suggested to determine the sample volume to be added to the cell.

Experiments Suggested

1. Determination of the titration end point based on titrating hydrochloric acid solution.
2. Adjustment of the end point of the titration in the cell.
3. Checking of the current efficiency of the reagent generation by titrating standard hydrochloric acid solution.
4. Determination of the concentration of unknown HCl sample solutions.

4.5. References

Bishop, E., Coulometric analysis, in *Comprehensive Analytical Chemistry,* Wilson, C. L. and Wilson, D. W., Eds., Elsevier, Amsterdam, 1975.

Chapter 5

Conductometric Techniques

Principle of the Technique

Conductometric techniques enable the determination of concentrations on the basis of the connection between the conductivity and the ion concentration of solutions.

In electrolyte solutions, electric conduction is brought about by the migration of ions. If an electric field is established between indifferent electrodes immersed into a solution containing ions, the latter will migrate in the direction to the electrode having a charge of the opposite sign than that of the ion. The conductivity of solutions is determined by the number and migration rate (mobility) of the ions. The conductivity of solutions is, at a given temperature, in a given solvent and at a given concentration a parameter characteristic of the system under test. Conductivity can be measured by a direct or by an indirect technique. In the case of direct measurement, the concentration of the tested component can be determined on the basis of the measured conductivity value only if the conductivity of other ions present in the solution is subtracted from the total conductivity. Accordingly, the quality and concentration of other ions has to be known.

"Indirect technique" means titration, which enables selective determination, provided that a titrating reagent reacting selectively with the component to be determined is applied. The selective reaction is the titration reaction, whose end-point is indicated by a change in the conductivity. Accordingly, in this latter technique it is not necessary to know the absolute value of the conductivity.

Conductivity can be measured by means of a DC or else by an AC current. DC measurements can be applied in practice only under adequate circumstances and consequently AC techniques are more widely used. According to the frequency of the AC applied, low-frequency (conductometric) and high-frequency (oscillometric) techniques can be distinguished.

5.1. Conductometry

Principle of the Technique _____

Conductometric measurements are carried out with low-frequency current, the frequency being lower than a few thousand hertz. In practice, the technique is used mostly for end-point indication in titrations (indirect conductometry). The technique enables acid-base, precipitation, redox, and complexometric titrations to be carried out. The end-point of the titration is defined by a break-point on the titration curve, in which changes in conductivity are plotted against the volume of the titrant added. Once the end-point of the titration has been established, the concentration to be determined can be calculated. In order to attain good accuracy, a break point as sharp as possible should appear on the titration curve at the end-point and accordingly it is preferable to always select the titrant with due consideration to the ionic mobilities.

In conductometric titrations the dipping ring electrode (the so-called "bell electrode") is extensively used; this is a pair of platinum electrodes, rigidly secured to a support and electrolytically plated with platinum of colloidal dispersion. Its modified version is the three-ring-type bell electrode which ensures a well-defined current path.

5.1.1. Determination of Hydrochloric and Acetic Acids. Determination of the Dissociation Constant of Acetic Acid

Principle of the Technique _____

By titrating hydrochloric and acetic acids, the conductometric titration curves of strong and weak acids can be compared, and the dissociation constant of the weak acid can also be calculated.

To the latter end, the following data are to be known:

1. The initial concentration of the weak acid, which is given by the result of the titration.
2. The effective conductivity of the weak acid at the beginning of the titration.
3. The conductivity of the solution of the completely dissociated weak acid (λ_∞).

All these data can be obtained from the different titration curves:

$$\lambda_{HOAC} = \lambda_{NaOAc} + \lambda_{HCl} - \lambda_{NaCl}$$

From the above data, on the basis of

$$\alpha = \frac{\lambda}{\lambda_\infty} \quad \text{and} \quad K_d = \frac{\alpha^2 c}{1 - \alpha}$$

the dissociation constant can be calculated.

Note: The two acids can be determined even when present together, but the accuracy is within 1 to 2% only within certain concentration limits. In the simultaneous determination of the two acids (see Gilbert's theory) the conductivity minimum causes a difficulty, therefore the optimum conditions under which this interference is small, are to be found. By reducing the dielectric constant of the medium, the simultaneous determination can be effected more accurately; therefore it is advantageous to make the simultaneous determination of HCl and CH_3COOH in ethyl alcohol-water mixture as a solvent. The accuracy of the determination of acetic acid at a concentration of 0.1 N in 50% ethyl alcohol can reach 0.1 to 0.2%. This is due to the reduction of the dielectric constant of the medium and a consequent reduction in the dissociation constant of acetic acid. Under these conditions, the conductivity minimum is shifted toward zero degrees of titration.

Apparatus

Pipette, 25 cm^3
Three volumetric flasks, 100 cm^3
Graduated cylinder, 100 cm^3
Beaker, 250 cm^3
Two burettes, 12 cm^3
Conductivity meter
Dipping ring electrode
Magnetic stirrer

Chemicals

0.2 M sodium hydroxide solution
0.2 M acetic acid solution

Procedure

1. Pipette 25.00 cm^3 of the hydrochloric acid solution into a 400 cm^3 beaker and dilute with 150 cm^3 of distilled water. Place the beaker on the magnetic stirrer,

immerse the electrode into the solution, start stirring, and adjust the stirring rate to a value where no vortex is formed. Adjust the instrument into measuring position and read the conductivity pertaining to zero volume of the standard solution added. Start the titration by adding the titrant, 0.2 M NaOH in 0.5 cm^3 portions, and read the conductivity after each addition.

2. Titrate a 25.00-cm^3 aliquot of the unknown acetic acid solution similarly, adding the titrant in 0.2-cm^3 portions up to 2 cm^3, then in 0.5-cm^3 portions up to 10 cm^3 of 0.2 M NaOH.

3. Titrate a 25.00-cm^3 aliquot of the unknown NaOH sample with 0.2 M CH$_3$COOH solution, adding the titrant in 0.5-cm^3 portions.

Evaluation

Plot the conductivity vs. the volume of titrant solution added, determine the equivalence point, and calculate the concentration of HCl, NaOH, and CH$_3$COOH, respectively, in g/100 cm^3 as well as the dissociation constant of acetic acid.

5.1.2.   Determination of Phenol

Principle of the Determination _____

Phenol can be determined as a weak acid by Gaslini and Nahum's method, by dissolving phenol in a weak base and titration with sodium hydroxide.

Apparatus

> Conductivity meter
> Dipping ring electrode
> Beaker, 150 cm^3
> Pipette, 5 cm^3
> Burette, 12 cm^3

Chemicals

> 0.1 N phenol solution
> 2 N ammonium hydroxide solution
> 0.1 N sodium hydroxide solution

Procedure

Draw an aliquot containing about 0.5 mmol of phenol from the unknown solution by a pipette, transfer it into a 150-cm^3 beaker, and dilute to 100 cm^3

with 2 N NH$_4$OH. Titrate the solution with 0.1 N NaOH, adding the titrant in 1-cm^3 portions, and read the conductivity after each addition. Continue the titration until three or five readings are obtained in the range where the conductivity increases steeply.

Evaluation

Plot the titration curve, find the equivalence point, and calculate the phenol concentration of the original solution in grams per liter.

5.2. Oscillometry

Principle of the Technique _____

The advantage of oscillometry over conductometry is that no galvanic contact is necessary between the solution and measuring device; thus the effect of polarization and other unwanted phenomena (catalysis, corrosion, etc.) can be eliminated. Closed systems, such as ampoules or heterogenous systems, can be measured directly.

In the high-frequency method, the titration cell is part of an oscillating circuit, and the changes in the high-frequency (>2 MHz) conductivity of the solution are measured by following the Q-factor or the resonance frequency of the oscillating circuit. The oscillotitrator Radelkis, Type OK-302 works with a capacitive-type cell, and measures the conductance by following the Q-factor of the circuit, by measuring the grid current of the oscillator tube.

The Q-factor of the circuit changes along a maximum-type curve with the concentration. Unequivocal measurement can only be carried out at concentrations where the changes during the titration remain within the ascending or descending section of the maximum-type curve. Therefore, it is necessary to know the high-frequency concentration curve of the solution examined.

5.2.1. Calibration and Sensitivity Curve of the Oscillotitrator for Hydrochloric Acid Solution

Apparatus

> Oscillotitrator
> Pipette, 1 cm^3 graduated
> Burette, 25 cm^3
> Beaker, 250 cm^3

Chemicals

1 N hydrochloric acid

Procedure

Pour 200 cm^3 of distilled water into a 250-cm^3 beaker, put in the stirring membrane, and fasten the beaker within the electrodes. The outside of the beaker is to be wiped completely dry prior to placing it within the electrodes. Care should be taken that the surface of the solution is above all three electrodes. Switch the instrument on, adjust appropriate sensitivity, start stirring, and add 0.1 cm^3 of 1 N hydrochloric acid and adjust the pointer of the instrument to the beginning of the scale. Then add 0.1-; 0.2-; 0.2-; 1.0-; 1.0-; 2.0-; 10.0-; and 20.0-cm^3 portions of 1 N hydrochloric acid, using a graduated 1-cm^3 pipette or a burette, and read the scale deflection after each addition.

Evaluation

Calculate the concentration of hydrochloric acid in gram-equivalents per liter, taking the dilution into consideration, and plot the calibration curve, i.e., scale deflection vs. the logarithm of concentration.

Plot also the sensitivity curve, which is the derivative of the calibration curve, and give the maximum sensitivity of the instrument for the cell used.

5.2.2.   Determination of Sulfuric Acid

Principle of the Determination _____

In the titration of a strong acid with a strong base, the conductivity of the solution decreases up to the equivalence point due to the reduction in H^+ concentration, then increases due to the increase in OH^- concentration.

Apparatus

Oscillotitrator
Beaker, 200 cm^2
Graduated cylinder, 100 cm^3
Burette, 12 cm^3

Chemicals

0.1 N sodium hydroxide standard solution

Procedure

Pipette an aliquot containing about 1 meq of sulfuric acid from the unknown solution into a-200 cm^3 beaker, and place the beaker within the electrodes. Place in the stirring membrane and fill up with distilled water until the level of the liquid is above the upper electrode. Adjust the appropriate sensitivity, start stirring, and adjust the pointer to maximum deflection. Then start the titration with 0.1 N NaOH, adding the titrant in 0.5-cm^3 portions and reading the scale deflection after each addition. Take four or five readings after the equivalence point.

Evaluation

Plot the titration curve, find the equivalence point, and calculate the sulfuric acid concentration of the unknown in mg/100 cm^3.

5.2.3. Determination of Acetic Acid with Methyl Glucamine

Principle of the Determination _____

The determination of a weak acid is preferably carried out by using a weak base as a titrant, as in this case the change in the slope of the titration curve is great at the equivalence point. Methyl glucamine (1-desoxy-1 methylamino sorbitol) is a very good titrant, its dissociation constant being close to that of ammonium hydroxide ($K_d = 1.5.10^{-5}$), its solution being stable, and the compound itself nonhygroscopic. A solution prepared by dissolving an accurately weighed amount in a known volume of water need not be standardized.

Apparatus

> Oscillotitrator
> Beaker, 200 cm^3
> Graduated cylinder, 100 cm^3
> Burette, 12 cm^3
> Pipette, 10 cm^3

Chemicals

> 0.1 N methyl glucamine solution
> 0.1 N acetic acid solution

Procedure

Bring a 10.00-cm³ aliquot of 0.1 N acetic acid into a 200-cm³ beaker using a pipette, and fill up the solution to about 150 cm³ with distilled water so that the surface of the liquid gets above the electrodes. Place the beaker within the electrodes, put in the stirrer, and start stirring. Titrate the solution by adding 0.1 N methyl glucamine in 0.25-cm³ portions, and reading the scale deflection after each addition.

Evaluation

Plot the titration curve, find the break point, and then calculate the acetic acid concentration of the original solution in mg/100 cm³.

5.2.4. Determination of Chloride in Tap Water

Principle of the Determination _____

On titrating the sample with a silver nitrate standard solution, chloride is precipitated as AgCl:

$$Cl^- + Ag^+ + NO_3^- = AgCl + NO_3^-$$

During the titration, chloride ions are replaced by nitrate ions, the latter having only somewhat lower mobility than the former. Thus the conductivity does not change remarkably during titration and, when small amounts of chloride are titrated, practically no change in conductivity is indicated by the instrument. After the equivalence point the conductivity increases, which is indicated by a reduction in the grid current.

Apparatus

> Oscillotitrator
> Beaker, 200 cm³
> Microburette

Chemicals

> 0.1 N silver nitrate standard solution

Procedure

Dilute 100 cm³ of tap water with distilled water so that the level of the liquid is higher than the uppermost electrode of the capacitive cell. Then titrate with

0.25-cm^3 portions of 0.1 N silver nitrate standard solution under stirring, taking readings after each addition and taking two or three points in the portion where the instrument shows a steep increase in conductivity.

Evaluation

Plot the titration curve and give the chloride concentration of the water sample in mg/100 cm^3.

5.2.5.   Oscillometric Examination of Ampoules

Principle of the Determination _____

The assay of liquids sealed in ampoules can be carried out with the help of a calibration curve made with standard solutions. The standard solutions are prepared with varying concentrations but with the same composition as that of the unknown sample. Naturally the concentration of the unknown sample must be within the concentration range of the standards.

   If the sample studied is, for example, a potassium chloride solution, the standards should also be made of potassium chloride in known concentrations, e.g., 1, 0.1, 0.05, 0.02, 0.01, 0.005, 0.002, and 0.001 M. The standards are poured and sealed into the ampoules, the conductivity of which is determined. On the basis of the measurements a calibration curve is prepared (meter deflections vs. concentrations of the potassium chloride standards), which is used for the evaluation of the concentration of the unknown sample.

Apparatus

> Oscillotitrator
> Ampoule-tester accessory
> Ampoules
> Pipettes, 2 and 5 cm^3
> Volumetric flask, 100 cm^3

Chemicals

> 1 N potassium chloride standard solution

Procedure

Place the ampoules containing the standard solutions and the unknown potassium chloride sample in the ampoule-tester accessory and read the deflections of the meter. Plot the calibration curve and determine the potassium chloride concentration of the unknown sample.

The positions of the switches "Sensitivity" and "Adjustment" should not be changed during the measurements of the standards and the sample.

5.3. References

Ewing, G. W., The measurement of electrolytic conductance, *J. Chem. Educ.,* 51, A469, 1974.

Pungor, E., *Oscillometry and Conductometry,* Pergamon Press, Oxford, 1965.

Stark, J. T., Two centuries of quantitative electrolytic conductivity, *Anal. Chem.,* 56, 561A, 1984.

Chapter 6

Dielectrometry

Principle of the Technique

The knowledge of the dielectric constant gives information on the properties of the material examined (e.g., about the structure, about the purity, or water content of the material).

The measurement of the dielectric constant can be carried out in various ways; the methods most often used are as follows:

1. Methods based on the measurement of AC capacity: bridge, resonance, or heterodyne method.
2. Methods based on phenomena connected with electromagnetic waves.

The determination of the dielectric constant of liquids is generally carried out with the help of a calibration curve. Measurement of the dielectric constant of powders is possible by the immersion method or by the direct method.

For a measuring cell the following equations are valid:

$$C_o + C_s = C_{total}$$

$$\varepsilon C_o + C_s = C_{material}$$

$$C_o = \frac{C_{material} - C_{total}}{\varepsilon - 1}$$

$$\varepsilon = \frac{C_{material} - C_s}{C_o}$$

where C_o is the basic capacity and C_s is the stray capacity.

6.1. Determination of the Water Content in Methyl Alcohol

Principle of the Determination _____

In general, the determination of the water content is the most useful area of dielectrometry. The water content of different mixtures, powders, and oils can be determined, using a calibration curve plotted on the basis of the results obtained in standards containing different amounts of water.

Apparatus

> Universal dielectrometer
> Dielectric cell of 1 pF basic capacity
> Volumetric flasks
> Pipettes

Chemicals

> Methyl alcohol (water content: 0.1%)

Procedure

Prepare methyl alcohol-water mixtures the alcohol content of which is 5, 20, 50, 70, 80, 90, and 100%. Measure the capacity of the cell containing the different mixtures and plot the calibration curve.

Evaluation

Determine the water content and the dielectric constant of the unknown sample.

Chapter 7

Biocatalytic Sensors

Introduction

Biosensors are dedicated to selectively determine different components in complex matrices. The selectivity of their function is provided by a highly specific biorecognition process. During biosensor operation the physicochemical changes induced by the recognition reaction are transducted to an appropriate signal. Many different versions of biosensors have been developed recently. Several reviews and monographs have appeared on the subject.

Biosensors that employ a biocatalytic process for recognition and an electroanalytical base-sensing element for signal transduction are often called enzyme electrodes. Enzyme electrodes can be considered as the most established, widely studied, and employed electroanalytical biosensors. They are made of an electroanalytical base-sensing element such as ion-selective electrode, amperometric electrode, etc. and a reaction layer containing an immobilized biocatalyst.

The biocatalyst employed is in most cases an isolated enzyme but a catalytic plant or animal tissue, isolated organelle suspension, or microorganism culture can also be used successfully.

For measurements with biocatalytic electrodes the flow injection arrangement is very advantageous.

In the following section three different laboratory experiments are described in which the students prepare different kinds of biocatalytic electrodes and study them in batch-type experiments.

The biosensors prepared according to the methods given here will be used in a flow injection experiment presented in the flow analysis section.

7.1. Glucose Determination in Fruit Juice with Amperometric Glucose Electrode

Principle

Glucose determinations are among the most often performed analytical measurements. Many analytical methods used for this are based on the reducing character of glucose molecules. However, the direct electrochemical oxidation of glucose on the surface of indifferent electrodes is very slow in the potential range accessible in aqueous media. This makes amperometric glucose measurement rather difficult.

The operation of the amperometric glucose electrode used in this laboratory experiment is based on the following enzymatic reaction:

$$\beta\text{-}D\text{-Glucose} + O_2 \xrightarrow{\text{glucose oxidase}} \text{gluconic acid} + H_2O_2$$

The glucose oxidase enzyme is a very stable and relatively inexpensive biocatalyst with good substrate selectivity. It recognizes glucose. Several alternatives exist for the electrochemical detection.

The decrease of oxygen concentration or the concentration of the produced hydrogen peroxide can be followed amperometrically. An acid is produced; thus, at low buffer capacity the pH change can also be used for indication. The hydrogen peroxide can be reacted further in different subsequent reactions producing changes detectable with an electrode. The peroxidase-catalyzed reaction between iodide ions and the hydrogen peroxide is a good example of this. Electroactive iodine is produced in the reaction. The loss of iodide activity can also be detected by an iodide-selective electrode. Recently the direct electron exchange reaction between horseradish peroxidase enzyme and different amperometric graphite electrodes has been exploited for the detection.

In this experiment the direct amperometric detection of hydrogen peroxide on platinum disc electrode is used.

Operation of the Enzyme Electrode

Glucose enzyme electrode is made by immobilizing glucose oxidase in a thin film reaction layer on the surface of a platinum disc electrode, and the enzyme electrode is placed in a measuring cell containing pH = 7.3 phosphate buffer, a silver/silver chloride reference, and a platinum wire counter electrode. Continuous stirring with a magnetic stirring bar is employed. Through the potentiostat, 0.7-V polarizing potential is applied to the electrode and the amperometric current is followed by recording on a strip chart recorder. At 0.7 V, the hydrogen peroxide undergoes electrochemical oxidation at the platinum electrode surface. Without glucose a very small current, the residual current, flows through the electrodes. This shows the absence of species electroactive at the potential of the working electrode. As a matter of fact, in the case of commercial glucose electrodes a special cellulose acetate membrane is employed on the platinum disc. The pores of this membrane allow the small hydrogen peroxide molecules to diffuse through selectively but exclude the larger molecules that may be present in the matrix.

When glucose is added to the solution, it diffuses into the reaction film, where it is converted to gluconic acid with the concomitant production of hydrogen peroxide as the equation shows. The reaction rate in certain conditions depends on the glucose concentration. If given conditions are provided a quasi-steady-state of the reaction and of diffusion processes is established. The local hydrogen peroxide concentration in the reaction layer will achieve a steady value that depends on the glucose concentration. On the platinum disc the following oxidation process will proceed:

$$H_2O_2 \longrightarrow O_2 + 2H^+ + 2e^-$$

The steady-state current from this electrode reaction will be proportional to the local quasi-steady-state concentration of the hydrogen peroxide generated by the enzymatic reaction. Thus the steady-state current — in certain conditions — is a direct measure of glucose concentration in the sample solution.

Apparatus

Potentiostat and connected current-time recording device
Platinum disc working electrode
Silver/silver chloride reference electrode
Platinum counter electrode
Magnetic stirrer
Measuring cell (it can be a 50- to 100-ml beaker)
Electrode holding device (a cork with three holes and a laboratory clamp)
Hamilton syringe

Chemicals

Phosphate buffer (pH = 7.3)
10 mmol/1 glucose standard solution (prepared 1 day before the experiment, so that the mutarotation is in equilibrium)
Glucose oxidase enzyme (EC 1.1.3.4.) commercial preparation or the immobilized reaction layer (film or polyacrylamide gel bar in supporting syringe)
Nylon hose tissue spacer
Dialysis membrane
Thread

Procedure: Preparation of the Enzyme Electrode

Method 1   (used when a reaction layer containing immobilized enzyme is provided)

The enzyme is entrapped in cross-linked polyacrylamide gel cured inside the cylinder of a plastic syringe, the end of which was removed. With careful movement of the plunger of the syringe about 0.1 to 0.3 mm length of the gel is pushed out. It is wetted with buffer and a thin disc is cut off. The gel bar is kept in the refrigerator at 4°C before and after use. The small disc is placed on the previously wetted end of the platinum working electrode. Care must be taken to avoid air bubbles between the electrode and the disc. For wetting, rinsing, and washing always buffer pH = 7.3 is used. The end of the electrode is covered with a cap of dialysis membrane. The dialysis film must be wet. Thread is used for fixing.

The procedure is slightly different if the reaction layer is provided in the form of a film. Then a cork drill can be used to cut out the properly sized disc.

Method 2   (used when soluble enzyme is provided in its commercial form)

As in the first step the previously washed and polished surface of the platinum disc base-sensing element is covered with a thin nylon net. It is made of nylon hose tissue stretched and tied with thread, winding the latter several times around the electrode body close to the end. This serves as a spacer keeping a constant reaction layer thickness. The extending part of the nylon tissue is cut off with a sharp razor blade. Glucose oxidase enzyme solution is made using the buffer as solvent. This is spread evenly on the electrode surface. After this, the end of the electrode is covered with a stretched dialysis membrane cap as mentioned before. Great care must be taken to avoid air bubble entrapment under the cap. The dialysis membrane cap is tied on by thread as before. The membrane cap immobilizes the enzyme by entrapping it, so it must not have any slits or holes. The extending part of the dialysis membrane is cut off and the electrode is ready for use. The dialysis membrane cap should be kept wet. When not in use the electrode should be stored in the refrigerator soaked in buffer solution. The well-made electrode can keep its glucose measuring activity quite long; however, it is advised to make a new electrode in every laboratory experiment.

A skilled person can easily prepare an electrode using as little as 5 µl enzyme solution. Therefore, it is enough if we prepare about 10 µl. It can easily be done if we place 5 mg enzyme on a hydrophobic surface (Teflon membrane, parafilm) and with a 10-µl Hamilton syringe employing repetitive suction and pressure we dissolve the enzyme in 10 µl buffer solution on the surface as a drop or liquid inside the syringe. The syringe is also used for applying the enzyme solution on the electrode. The enzyme solution can be stored in it; however, clogging of the syringe may happen. It is advised to wash out the solution after performing the laboratory experiments.

Procedure

Transfer 20 ml buffer solution to the measuring cell and immerse the electrodes. The amperometric measuring device is connected, 0.7 V polarizing potential is adjusted, and intensive, steady stirring is employed. The current is followed. When a steady-state background current is achieved 50- to 1000-μl volume of the standard or the sample solution is added into the cell. The current increases and in about 60 to 120 s achieves a steady-state value. The difference between this and the background current is the signal.

After the measurement the electrode and the cell are washed and new buffer is employed for the next measurement.

It is important to select the right current sensitivity range on the measuring apparatus. To be able to select this value a preliminary measurement is carried out with a standard solution in which we test the electrode activity.

Experiments Suggested

1. Check with a "naked" platinum disc working electrode (without the enzyme) if it gives amperometric signal to glucose concentration step.
2. Using the platinum electrode as a base-sensing element prepare the glucose enzyme electrode.
3. Study the glucose-measuring function of the amperometric electrode (preliminary measurement; current constant and response time are calculated).
4. Calibrate the electrode with glucose standard solutions (calibration curve is calculated and prepared).
5. Determine the unknown glucose concentration based on the calibration and measurements in sample solutions.

7.2. Urea Determination with Potentiometric Urea Biosensor

Principle

The determination of urea (or, as it is called, urea nitrogen) is a frequent task of clinical laboratories. Urea concentration measurements have importance in agricultural, environmental, and industrial analysis, too.

Many widely used analytical methods are based on the hydrolysis of urea and the detection of the product of this reaction. The reaction of hydrolytic decomposition of urea has a selective biocatalyst, the urease enzyme. This enzyme is one of the best known and most widely studied biocatalysts. It can be obtained from different beans and other sources and can be isolated in the form of highly active and stable preparations. It is not accidental that the first potentiometric enzyme electrode ever worked out served for urea measurements

and was based on the enzymatic hydrolysis of this substrate. In this laboratory experiment the operation of an advanced version of this electrode is
studied.

The basis of the enzyme electrode function is the selective recognition of
urea by the urease enzyme. The following reaction takes place:

$$\text{Urea} + 3H_2O \xrightarrow{\text{urease}} 2\,NH_4^+ + HCO_3^- + OH^-$$

This reaction can be followed using different potentiometric electrodes. The
activity of the liberated ammonium ions can be measured with an ion-selective
ammonium electrode. Ammonia gas-measuring potentiometric cells are also
applicable if a high pH of the medium is adjusted. Potentiometric carbon
dioxide gas probes as well as pH electrodes can also be considered for application
here.

In this experiment a neutral carrier (nonactin)-based ion selective ammonium electrode is used as a base-sensing element. The measuring surface of this
electrode is coated with a thin film reaction layer. This contains immobilized
urease enzyme. Urea measurements are performed in vigorously stirred Tris
buffer with the urea enzyme electrode prepared in this way. (The selectivity of
the ammonium ion electrode does not allow to use buffer solutions that contain
potassium ions in high concentration.)

Silver/silver-chloride reference electrode is connected to the measurement cell through a lithium acetate current bridge. The potential of the urea
electrode is followed in time using a high-impedance millivolt meter. A strip
chart recorder or electronic recording unit is connected to the output of the
millivolt meter. Without urea the electrode potential will be determined by
the ion activities of the background buffer solution. When urea is added to
the stirred solution it will soon enter the reaction layer where it decomposes
upon the catalytic action of the immobilized enzyme according to the equation. The local activity of the ammonium ions formed is detected by the
electrode. The ammonium ion activity depends on the reaction rate. The
reaction rate, on the other hand, depends on the urea concentration (see
Michaelis-Menten kinetics). Under certain conditions the diffusion mass
transport and the reaction rate achieve a quasi-steady-state. This gives a
steady electrode potential which is the measure of the urea concentration.
The electrode potential value measured can be converted to urea concentration using an appropriate calibration or employing standard addition measuring technique.

It must be mentioned here that, if the sample solution contains ammonium
ions, or ions interfering with the base electrode function, the electrode will
detect them and we will get false urea concentration data. However, the
presence and extent of this kind of interference can be checked easily doing
potentiometric measurements in the sample with "naked", reaction-layer-less
ammonium ion-selective electrode.

Apparatus

High input impedance pH-mV meter
Potential-time recording unit
Magnetic stirring apparatus
Ion-selective ammonium electrode (PVC electrode body; flat, plasticized PVC measuring membrane containing nonactin as electrode active material)
Double junction silver/silver chloride reference electrode (with lithium acetate bridge electrolyte)
Measuring cell (it can be a 50- to 100-ml beaker)
Electrode holder (e.g., cork with holes for the electrodes and laboratory clamp)
Hamilton syringe

Chemicals

Tris buffer 0.05 to 1 mol/l, pH = 7.0 (it must not contain potassium ions)
1 mol/l urea standard solution (freshly prepared with 0.05 to 0.1 mol/l Tris buffer)
Urease enzyme (EC 3.5.1.5.) isolated commercial preparation or immobilized reaction layer (the latter can be enzyme entrapped in polyacrylamide gel bar in a supporting syringe)
Nylon hose tissue spacer
Dialysis membrane
Thread

Procedure: Preparation of the Enzyme Electrode

Method 1   (used when a reaction layer containing immobilized enzyme is provided)

The enzyme is entrapped in cross-linked polyacrylamide gel cured inside the cylinder of a plastic syringe with cutoff end. With careful movement of the plunger of the syringe the gel is pushed out to a length of about 0.1 to 0.3 mm. It is wetted with buffer and a thin disc is cut out. The gel bar is kept in the refrigerator at 4°C before and after use. The small disc is placed on the previously wetted end of the ion-selective ammonium electrode. Care must be taken to avoid air bubbles between the electrode and the disc. For wetting, rinsing, and washing the buffer (Tris buffer, pH = 7.0) always is used. The end of the electrode is covered with a cap of dialysis membrane. The dialysis film must be wet. Thread is used for fixation (see Figure 7.1). (Two students can easily make this work, one holding the electrode, the other tightening the thread.)

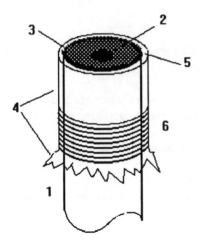

1-Electrode body
2-Platinum base sensing disc
3-Biocatalytic reaction layer
4-Dialysis membrane cap
5-Spacer
6-Thread

FIGURE 7.1
Biocatalytic electrode.

The procedure is slightly different if the reaction layer is provided in the form of a film. Then a cork drill can be used to cut out the properly sized disc.

*Method 2   (used when soluble enzyme is provided in its
              commercial form)*

As the first step the previously washed ammonium ion measuring disc base-sensing element is coated by a thin nylon net. It is made of nylon hose tissue stretched and tied with thread winding the latter several times around the electrode body close to the end. This serves as a spacer keeping a constant reaction layer thickness. The extending part of the nylon tissue is cut off with a sharp razor blade. Urease enzyme solution is made using the buffer as solvent. This is spread evenly over the electrode surface. After this the end of the electrode is covered with a stretched-on dialysis membrane cap as described before. Great care must be taken to avoid air bubble entrapment under the cup. The dialysis membrane cup is tied on by thread as described before. The membrane cup immobilizes the enzyme by entrapping it, so it must not have any slit or hole. The extending part of the dialysis membrane is cut off and the electrode is ready for use. The dialysis membrane cup should be kept wet. When not in use, the electrode should be soaked in buffer solution and stored in a refrigerator. The well-made electrode can keep its urea measuring activity quite long; however, it is advisable to make a new electrode in every laboratory experiment.

A skilled person easily can prepare an electrode using as little as 5 µl enzyme solution volume. It is enough if we prepare just about 10 µl. It can be done easily if we place 5 to 10 mg enzyme on a hydrophobic surface (Teflon membrane, parafilm) and with a 10-µl Hamilton syringe employing repetitive suction and pressure we dissolve the enzyme in 10 µl buffer solution on the surface as a drop or liquid inside the syringe. The syringe is also used for applying the enzyme solution on the electrode. The enzyme solution can be stored in it; however, clogging of the syringe may happen. It is advised to wash out the syringe after performing the laboratory experiments.

Procedure

Transfer 20.00 ml of the sample or standard solution into the measurement cell and adjust the magnetic stirrer to a steady speed. Introduce the urea electrode and the bridge of the reference and record the potential difference. A steady value of the cell voltage is taken as signal. Empty the cell, rinse the electrode thoroughly, and fill in the new solution for the next measurement. One measurement takes about 120 to 180 s, depending on the reaction layer thickness. When the electrode is placed from a very concentrated solution into a very dilute one (e.g., the concentration difference is about one order of magnitude in favor of the first solution), then the electrode must be soaked for 3 to 5 min in buffer solution between measurements to avoid cross-contamination.

Experiments Suggested

1. Calibration of the "naked" ammonium electrode with ammonium chloride standard solutions or with urea solutions (we shall observe that the base sensor does not detect urea; the urea solution must be freshly made!).
2. Preparation of urea enzyme electrode using the ammonium electrode as base sensing element.
3. Study the response of the urea electrode. The response time is determined from the recording obtained after a sudden single concentration step.
4. Calibration of the electrode with urea standard solutions (the calibration curve is made and its shape is discussed).
5. The unknown urea concentration is determined using the calibration curve and measured electrode potential values.

7.3. Microbiological Electrodes: Preparation and Study of a Potentiometric Yeast Electrode

Theory

As mentioned in the introduction, an intact microorganism culture can also be used as a selective molecule recognizing a functional part in biosensors. The structure of these microbiological sensors is very similar to that of the enzyme electrodes. The only difference is that the active part of the reaction layer here is a culture of living microorganisms, which, in the presence of the measurable species, produces a local concentration change detected by the base-sensing element. Materials used as substrate, as nitrogen, or energy sources by the culture can be determined in this way. In addition to this, heavy metals, mutagenic materials affecting the life or metabolic activity of the culture in very low concentration, can also be detected.

In research many different specially improved and refined strains were employed successfully to develop sensors of special function. In this laboratory experiment we study the function of microbiological sensors using a potentiometric yeast electrode. *Saccharomyces cerevisiae,* i.e., Baker's yeast, is readily available. It is easy to keep alive and it has high metabolic activity. In the presence of certain saccharides such as glucose or sucrose, as a consequence of the fast aerobic metabolism in the culture, detectable pH changes may occur. This can be used to make yeast-based saccharide electrode.

A potentiometric pH electrode is used as base-sensing element. A glass-shielded antimony micro-disc electrode or flat-end cylindrical glass electrode can be used for this. We cover the surface of this with a layer of yeast suspended in low buffer capacity phosphate buffer. The yeast cells are retained on the electrode surface by a dialysis membrane cap. In sugar-free solution the metabolic activity in the cell is small. Therefore, the pH at the electrode surface will be the same as in the bulk of the solution. In the presence of certain sugars, however, the microbial metabolism produces a pH difference between the yeast layer and the bulk. This difference is detected by the pH electrode. In a given range the pH difference depends on the sugar concentration. In this case the change in the potential of the pH-sensitive base sensor is the measure of the sugar concentration.

The pH changes to lower values with increasing metabolic activity. Therefore, to achieve a large signal the application of a buffer of higher pH (pH = 8) is advantageous. In industrial fermentation with the yeast the pH is kept between 3 and 5 to avoid bacterial infection.

To select the right buffer for the measurement, we need to consider that the capacity of the buffer influences the signal. High buffer capacity can quench the local pH difference. For our measurements, buffer with low capacity is needed.

Note that the yeast-based biosensor studied in this experiment has higher didactic than practical value. Its practical disadvantages are as follows: The response of this electrode is not selective and it is influenced by the sample pH as well as by the sample buffer capacity.

Apparatus

> High-impedance mV meter
> Potential-time recording unit and/or computerized data collection
> > unit

Magnetic stirrer
pH-sensitive potentiometric electrode with flat cylindrical
measuring surface, (glass electrode or shielded antimony disc
electrode)
Silver/silver chloride reference electrode
Measuring cell (50- to 100-ml beaker)
Electrode holder (e.g., cork with holes for the electrodes and
laboratory clamp)

Chemicals

Sucrose solution 1 mol/l (freshly prepared)
Glucose solution 0.5 mol/l (1 day old)
0.5 mol/l phosphate (pH = 8) buffer containing sodium chloride
in physiological concentration
Baker's yeast *(S. cerevisiae)* obtained from any food store;
(suspension made 1 day before)
Dialysis membrane
Thread
Nylon hose tissue spacer

Procedure: Preparation of the Potentiometric Microbial Sensor

The previously washed surface of the potentiometric base-sensing element is
coated with a thin nylon net. It is made of nylon hose tissue stretched and tied
with thread winding it several times around the electrode body close to the end.
This serves as a spacer keeping a constant reaction layer thickness. The extending
part of the nylon tissue is cut off with a sharp razor blade. A thick suspension of
yeast cells is placed on the measuring surface using a Pasteur pipette. The
suspension is made with pH = 5 phosphate buffer containing sodium chloride in
physiological concentration. It is spread evenly on the electrode surface. After
this, the end of the electrode is covered with a stretched dialysis membrane cap
as described before. Great care must be taken to avoid air bubble entrapment
under the cap. The dialysis membrane cap is tied on by thread as described
before. The membrane cap immobilizes the cells by entrapping them, so it must
not have any slit or hole. The extending part of the dialysis membrane is cut off
and the electrode is ready for use. The dialysis membrane cap should be kept wet.
When not in use the electrode should be stored in the refrigerator soaked in buffer
solution. The well-made electrode can keep its measuring activity quite long;
however, it is advised to make a new electrode in every laboratory experiment.

Measurements

Transfer 20 ml of the very dilute phosphate buffer solution into the measurement cell and adjust the magnetic stirrer to a steady speed. Immerse the yeast

electrode and the reference electrode and record the potential difference. After a steady value is established, add 10 to 500 µl of the sample or standard solution and record the cell voltage change. When a steady value is obtained the difference of this and the one measured in the buffer is taken as a signal. Empty the cell, rinse the electrode thoroughly, and introduce new solution for the next measurement. One measurement takes about 120 to 240 s, depending on the reaction layer thickness. When the electrode is placed from a very concentrated solution into a very dilute one (e.g., the concentration difference is about one order of magnitude in favor of the first solution), the electrode must be soaked for 3 to 5 min in buffer solution between measurements to avoid cross contamination. When handling the electrode it needs to be kept in mind that a living culture provides the measurement signal. We should always keep the electrode in buffer solution and use buffer for washing to avoid osmotic shock. The pH = 8 we use in the measurement is not optimal for yeast. For storing the suspension and the electrode higher buffer capacity and lower pH is recommended (phosphate buffer of pH 4 to 6).

Experiments Suggested

1. It is checked if the biocatalyst-free base sensor responds to glucose or sucrose standards.
2. The yeast electrode is prepared using the pH-sensitive electrode as base-sensing element.
3. The response of the microbial electrode is studied. The response time is determined from a recording obtained after a sudden single concentration step for glucose and sucrose. The time needed to reach 95% of the total potential change is used to characterize the response time of the electrode.
4. The electrode is calibrated with glucose and sucrose standard solutions (the calibration curve is constructed and its shape is discussed).
5. The unknown sucrose concentration is determined using the calibration curve and the measured electrode potential change values.

7.4. References

Hall, E. A. H., *Biosensors,* Open University Press, Milton Keynes, 1990.

Scheller, F. and Schubert, F., *Biosensoren,* Akademie-Verlag, Berlin, 1989.

Part II

Optical Analytical Techniques

Chapter 8

Spectrophotometry

Principle of the Technique

The subject of spectrophotometry is to study interactions between electromagnetic radiation and matter. If an electromagnetic radiation acts upon a piece of matter and there is selective absorption, the material will absorb different amounts of the components of the radiation of different wavelengths. The change in the degree of absorption as a function of wavelength is the absorption spectrum. The latter is characteristic of the quality of the material. The degree of absorption of light of a given wavelength enables a conclusion to be drawn on the amount of the material under test. The basis of quantitative determination is Beer's law:

$$A = \lg \frac{I_0}{I} = \varepsilon \, cl$$

where A is the light absorption (absorbance) of the material under test, I_0 is the light intensity without absorption, I is the intensity of light which has passed through and emerges from the solution containing the material to be tested, l is the layer thickness of the solution containing the material to be tested (absorption path length), ε is the molecular absorption coefficient (with the concentration given in moles $\times \, 1^{-1}$ and 1 in cm units), and c is the concentration of the material under test.

Instead of absorbance it is possible to measure transmittance (T):

$$T = \frac{I}{I_0} \quad \text{or} \quad T\% = 100 \frac{I}{I_0}$$

According to the wavelength used for the measurement, absorption spectrophotometry can be ultraviolet (UV), visible (VIS), and infrared (IR) spectrophotometry. The VIS and UV spectrophotometric techniques are used mainly
for the quantitative determination of components present at low concentration
levels. IR spectrophotometry also enables quantitative determinations on the
basis of Beer's law; however, it can be used only for the determination of
principal components. The IR measuring technique is mainly used for qualitative analysis. By application of the characteristic bond and group frequencies,
IR spectrophotometry can be used for the identification of gases, liquids, and
solids, for molecular structural analysis, for checking the purity of substances,
etc. On the basis of the position and intensity of a few characteristic bands or
knowing the whole spectrum, UV and VIS spectrophotometry can also be used
for qualitative analysis; however, these spectra are much less characteristic of
the substances under test than are IR spectra.

8.1. Application of IR Spectroscopy in Qualitative Analysis and Structure Elucidation: Evaluation of an Infrared Spectrum, Functional Group Analysis, and Structural Analysis

Principle of the Determination _____

The frequencies of the normal vibrations of molecules, i.e., the position of the
spectrum bands obtained (expressed in wavelengths or in wave numbers) are
determined by the masses of the atoms of the molecule and the forces acting
between the masses and consequently infrared spectra are individual to a high
degree. The slightest difference in structure brings about a variation in the
spectrum.

The two main fields of application of infrared spectroscopy are structural
analysis and qualitative analysis. In qualitative analysis, the presence of a
compound or of compounds of known structure in an unknown sample is to be
detected. Alternatively, the bonds and groups present in the compound are to
be determined on the basis of their infrared spectra or of spectra superimposed
upon one another. In structural analysis, the structure of a molecule newly
synthetized or of a hitherto unknown structure is to be established. In such
cases, first the characteristic functional groups are identified, and afterward
their configuration, the symmetry of the molecule, the distances between the
atoms building up the molecule, and the valency angles and the force constants
corresponding to the individual bonds are determined. Tasks of this kind could
be solved up to now only in the case of simple molecules on account of the
mathematical difficulties involved and as yet unsolved.

No strict rules and orders can be defined on qualitative analysis and structural analysis. The approach to the problem depends on the nature of the task and may involve subjective aspects. The methodology of the solution of tasks belonging to these two groups is different; however, there are many common characteristics as well.

Apparatus

Tables for the characteristic bond and group frequencies (Figures 8.1, 8.2, 8.3, and 8.4)

Tables compiled on the basis of the works of Colthup, Ullmann, Jones, Sándorfy and Cross, Sadtler, and DMS (Dokumentation der Molekül-Spektroskopie) spectrum catalogs

Procedure

In the following, the most general rules pertaining to the evaluation of infrared spectra from a qualitative analytical point of view will be presented. The solution possibilities of the problems of structural analysis will be dealt with shortly.

Qualitative analytical tasks may be of varied nature. In the simplest case, the spectra of the supposed compound or compounds are at our disposal and the presence of the individual substances can be detected by a simple comparison.

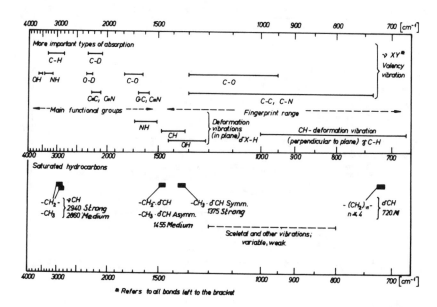

FIGURE 8.1

Diagram of characteristic bond and group frequencies according to Cross.

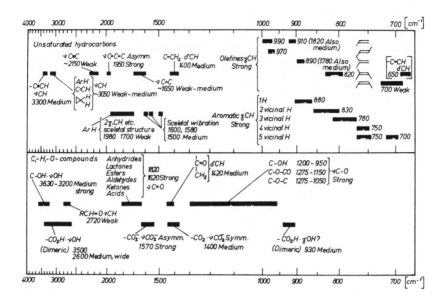

FIGURE 8.2
Diagram of characteristic bond and group frequencies according to Cross.

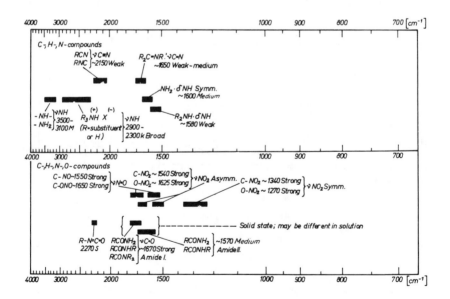

FIGURE 8.3
Diagram of characteristic bond and group frequencies according to Cross.

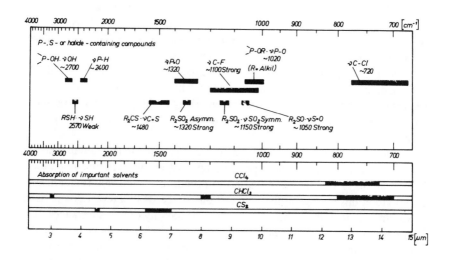

FIGURE 8.4

Diagram of characteristic bond and group frequencies according to Cross.

However, most often absolutely nothing is known of the sample to be tested. In this case it is preferable to start with a qualitative and quantitative elemental analysis. The empirical formula and the characteristic absorption bands appearing in the infrared spectrum enable to identify the bond types and functional groups present in the molecule. The characteristic bond and group frequencies were determined on the basis of spectra obtained with a great number of compounds of known compositions and accordingly the results obtained as described in the foregoing are of approximate nature, though generally they are correct. However, it is advisable to verify our conclusions by chemical or other techniques.

Evaluation of the Spectra

The work is started by assigning the bands of high and medium intensities.

The first step is to study the range corresponding to the OH, NH, and CH bond vibrations (3600 to 2800 cm^{-1}). This range enables a decision on the presence or absence of OH and NH groups (problems arise only if the OH or NH band merges into the baseline). These two groups can generally be distinguished. Aliphatic and aromatic CH groups can also be recognized and distinguished in most cases. In this case, the low intensity of aromatic bond vibrations presents a problem since these bands often merge into another band of higher intensity. Similarly, the CH bond vibrations of unsaturated groups overlap the aromatic CH bond vibration range and consequently saturated, unsaturated, or aromatic CH groups can actually be defined by deformation vibrations.

2850 to 1850 cm⁻¹. Not many bands are found in the spectra in this range. Carboxyl groups, hydrochlorides, SH groups, triple, and cumulative double bonds and α-amino acids cause absorption in this range. In addition to these, a few combination bands or harmonics may be observed.

1850 to 1470 cm⁻¹. Double bonds, aromatic skeletal vibrations, and NH deformation vibrations absorb in this range. The supplementary bands of the groups supposed on the basis of bands appearing in this range (e.g., carbonyl, aromatic, amine, nitro groups, etc.) should be checked in the lower higher wave number ranges.

1500 to 650 cm⁻¹. Spectra are the most complicated in this range. Identification of all of the bands is impossible and not even advisable, since their interpretation is unreliable. A varied multitude of bond, deformation, and group vibrations appear in this range. If one succeeds in a direct comparison of spectra, the identity of this range means the identity of two substances. It is justified that this range is called "fingerprint range"; the frequencies of the bands appearing in this range are controlled by other parts of the molecule to a higher degree and consequently they are more characteristic of the molecule in question than the range of higher wave numbers. In the latter, the appearing bands can generally be assigned to some particular sort of bond whereas other parts of the molecule have little influence upon them.

Having the origin of the bands of high and medium intensities more or less elucidated, the next step is to study the low-intensity bands. This is a much more difficult task, since the multitude of combination bands and harmonics produce low intensities and can be distinguished only with difficulty or cannot be distinguished at all from the similarly low-intensity basic bands.

It should be borne in mind that conjugation, or formation of association structures, may exert a great influence upon the position of the bands. The appearance of a band at the expected position is not always proof for the presence of the supposed group. In order to be able to accept its presence, other characteristic bands of that particular group should be looked up in the spectrum.

The absence of a band is not always a negative proof, since strong interactions (for example, the formation of a chelate or of some other structure characterized by a strong hydrogen bridge) are apt to shift some bands to a considerable degree or cause them to merge into the baseline (OH and NH bands).

On the basis of the spectrum, first some possible structures are supposed; from among these, the real one — or that most approximating it — is chosen on the basis of the empirical formula or of some other physicochemical techniques.

A helpful tool in the establishment of the structure is the examination of various derivatives or decomposition products. As a result of hydrogenation, deuterization, esterification, methylation, salt formation, acylation, etc., some absorption bands alter their position, shape, intensity, or number. A number of conclusions on the original groups can be drawn from these changes. Frequently it is also helpful to examine the spectrum of a substance in different states, such as solid or liquid form, solutions made with different solvents and at different concentrations, spectra recorded in nujol (suspension in paraffin oil). Conclusions on the intramolecular and intermolecular connections can be drawn from the changes in the spectrum. The shift in frequency observed while applying different solvents also gives information on the nature of the group in question.

When making an assignation, the intensity values of the bands should also be taken into consideration.

The intensity values can advantageously be made use of in the case of overlapping bands for the sure recognition of various groups. In such cases, the decision can be made either on the basis of data taken from the literature or else by measuring the intensities of the bands of standard model substances.

The most perfect solution of the task is to carry out a complete vibration analysis for the total molecule by application of computers. Of course, this is only possible in the case of simpler molecules of a high order of symmetry, even with application of the most up-to-date computer.

Finally it should be noted that simple mechanical evaluation methods are also known. For this purpose, punched cards are used and from these, with a little bit of luck, the spectrum identical to that of the sample under test can easily be selected by aid of sorting devices.

It is apparent from the aforesaid that there exists no general procedure for the evaluation of spectra. The auxiliary methods that are at disposal, the nature of the problem itself, and the skill of the analyst are of decisive influence upon the manner and the very success of the solution of the task.

Although, as we have seen, the evaluation of the spectra is always an individual task, the following order of action is recommended as an orientation: list the positions of the bands appearing in the spectrum, starting with the high wave numbers and proceeding toward the smaller ones. Hereupon, from appropriate tables identify the characteristic bonds or groups corresponding to the individual wave numbers, and write them next to the wave number.

The table contains the characteristic bond and group frequencies.

Evaluation

1. On the basis of bands observed in the spectrum and identified by aid of the tables construct the structural formula considered most probable. Do not take into consideration any groups unless all of their characteristic absorption peaks have been found in the spectrum and unless these are in agreement with the empirical formula.
2. Having carried out the above evaluation, the student is given the correct structural formula. Knowing the exact formula, the assignment of bands is carried out in such a manner that the symbol of the corresponding groups is written next to the wave numbers of the absorption peaks and a note defining the type of vibration is also made.

On the basis of the above studies, the following table is to be prepared:

Wave number of absorption peaks (cm^{-1})	Supposed groups	Groups present and their types of vibration

The first two columns should be filled out while studying the spectrum, the third one only when the exact structural formula is already known.

8.2. Application of IR, UV-VIS, and NIR Spectroscopy in Quantitative Analysis: Quantitative Analysis of Multicomponent Systems

Principle of the Determination

The possibility for the quantitative analytical application of the IR spectropho-tometric technique is based on the fact that the IR spectrum is a specific characteristic of the substances, i.e., there are differences even in the case of highly similar substances and that the spectra of multicomponent systems are built up additively of the spectra of the components. Up to three or four components, the technique enables simultaneous, nondestructive, speedy, and simple determinations to be carried out. Above this number of components it is necessary to carry out a separation prior to the measurement. The accuracy of the determination of the principal components is generally 1 to 2%; this error can in special cases be decreased.

Beer's law is also valid in the infrared range; however, when applying it, deviations may occur on account of the finite slit width and interactions between the components. In order to carry out quantitative analysis, absorption bands should be found for every component which are characteristic of the component in question, i.e., they do not coincide with the bands of the solvent or of other components of the mixture.

In elaborating a procedure for a quantitative determination, generally two techniques are applicable:

1. If the selected absorption bands do not overlap, the function absorbance vs. concentration is determined separately for each band on the basis of Beer's law by aid of a calibration series (see exercise 8.2.1. (A) and 8.2.1. (B), etc.).
2. If the bands overlap, it is necessary to record the IR spectra of all the components as well as that of the mixture of unknown composition and the concentrations are calculated by solving a set of equations. The latter is constructed in the following manner: in the case of n components, n "analytical positions" are selected in such a manner that at each position it is another component whose absorption is the dominant one. The absorption coefficient

values are determined at each position for each component and in this manner n^2 absorption coefficient values are obtained. The absorbance values at the selected n positions are determined in the spectrum of the unknown sample; it is possible to write n equations for these values. The terms of these equations are the absorbance ratios caused by the individual components. These are obtained as the products of the absorption coefficients previously determined, the layer thickness, and the concentrations.

$$A_1 = \sum_{i=n}^{n} \varepsilon_{1i} \, c_i \, l$$

$$A_2 = \sum_{i=n}^{n} \varepsilon_{2i} \, c_i \, l$$

$$A_n = \sum_{i=n}^{n} \varepsilon_{ni} \, c_i \, l$$

where A_1, A_2, A_n are the absorbance values at the selected positions; ε_{1i}, ε_{2i}, ε_{ni} are the absorption coefficient values at the selected positions (positions 1, 2, ... n; i components, 1 to n); c_i are the unknown concentrations; and l are the layer thicknesses.

By solving the set of equations, containing n members and n unknowns, the unknown c_i concentrations can be determined (see exercise 8.2.2.).

8.2.1(A).   Determination of 2,4-Tolylene Diamine and 2,4-Tolylene Diurea in the Presence of Urea, With Baseline Method

Principle of the Determination _____

The procedure enables following the preparation reaction of 2,4-tolylene diurea, a compound used as a raw material in the plastics industry as well as for checking the purity of the product obtained.
    The preparation reaction is the following:

| 2,4-tolylene diamine | 2,4-iso-cyanato toluene | 2,4-tolylene-diurea |

Other components may also be produced in the reaction; however, by adjusting the parameters it can be attained that the product contains only three components; the principal component (2,4-tolylene diurea) containing, as impurities, unreacted raw material (2,4-tolylene diamine) and a side product, urea.

The amount of 2,4-tolylene diurea is a measure of the progress of the reaction and of the purity of the product.

Quantitative determination according to the present procedure is based on the fact that it was possible to find bands characteristic of 2,4-tolylene diamine and 2,4-tolylene diurea, respectively, which coincide neither with each other nor with bands of the other component. These are the following: at 855 cm^{-1}, a sharp band produced by the torsion (γ as NH$_2$) vibration of NH$_2$ groups is characteristic of 2,4-tolylene diamine; at 1365 cm^{-1} a band of high intensity, produced by the symmetrical deformation vibration (δ_s CH$_3$) of the CH$_3$ group is characteristic of 2,4-tolylene diurea. (In the calculations of the evaluation, the positions selected according to the above were marked v_2.)

It is not necessary to select an analytical position for urea since its amount is the part missing from 100%.

Apparatus

        Infrared spectrophotometer
        Agate mortar or vibrator
        Hydraulic press with accessories
        Analytical balance
        Precision balance

Chemicals

        2,4-tolylene diamine (20 mg)
        2,4-tolylene diurea (20 mg)
        Potassium bromide (15 g)

Procedure

In order to produce the calibration standard series, "stock mixtures" are prepared in the following manner:

    A: 20 mg 2,4-tolylene diamine/5.00 g potassium bromide
    B: 20 mg 2,4-tolylene diurea/5.00 g potassium bromide
    C: 20 mg urea/5.00 g potassium bromide

Mixtures A, B, and C are prepared in such a manner that the material weighed on an analytical balance is placed into an agate mortar or a vibrator and the prescribed amount of potassium bromide (5 g) is added in three portions and mixed thoroughly.

The following calibration series is prepared from the stock mixtures described in the foregoing:

No.	Weighed in (g)		
	A	B	C
1	0.80	0.00	0.20
2	0.60	0.20	0.20
3	0.50	0.30	0.20
4	0.40	0.40	0.20
5	0.30	0.50	0.20
6	0.20	0.60	0.20
7	0.00	0.80	0.20

The components, weighed in on a precision balance, are mixed thoroughly in an agate mortar or vibrator, placed into the die of the tablet press, and pressed at a pressure of 250 at for 2 min. Hereupon the tablets are placed into the left-hand sample holder and their spectra are recorded, against a pure potassium bromide tablet in the reference beam, by means of an IR spectrophotometer model spektromom 2000.

The spectra are recorded in the 800- to 900-cm^{-1} and 1300- to 1400-cm^{-1} ranges.

The unknown samples to be analyzed are prepared according to the aforesaid and their spectra are recorded.

Evaluation

The evaluation is carried out with the baseline technique illustrated in Figure 8.5. The condition of the applicability of this technique is that the absorption of the impurity is a linear function of wavelength in the measurement range.

In addition to the analytical positions (v_2) selected on the basis of the spectra of the pure components — in our case these are 855 and 1365 cm^{-1} — two further measurement points (v_1 and v_3) are established on the right and left side at equal distances. The absorbance values are read from the spectrum at these points (A_1, A_2, and A_3). Should the instrument be such that it records transmittance values, the absorbance values can be obtained from the τ_1, τ_2, and τ_3 values by the following formula:

$$A = -\log \tau$$

Since positions v_1 and v_3 were selected as equidistant and the absorbance of the substance other than that to be determined varies linearly in the v_1 to v_3 range,

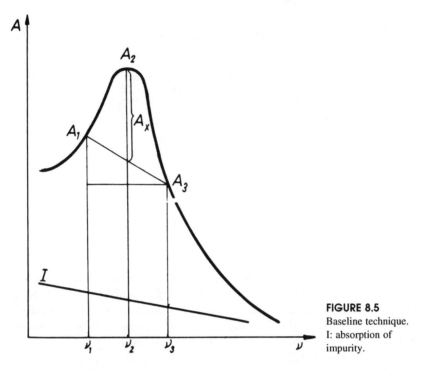

FIGURE 8.5
Baseline technique.
I: absorption of
impurity.

the absorbance values used to establish the function concentration vs. absorbance can be calculated with the following formula:

$$A = A_{v2} - 0.5\,A_{v1} - 0.5\,A_{v3}$$

The results of the calculations are summarized in a table:

	Concentrations (%)			Measurement positions (v_1, v_2, v_3) (cm^{-1})						
No.	Diamine	Diurea	Urea	τ_{v1}	τ_{v2}	τ_{v3}	A_{v1}	A_{v2}	A_{v3}	$A_{855\,cm} - 1 = A_{v2} - \dfrac{A_{v1}}{2} - \dfrac{A_{v3}}{2}$
1	80	0	20							
2	60	20	20							
3	50	30	20							
4	40	40	20							
5	30	50	20							
6	20	60	20							
7	0	80	20							
				τ_{v1}	τ_{v2}	τ_{v3}	A_{v1}	A_{v2}	A_{v3}	$A_{1365\,cm} - 1 = A_{v2} - \dfrac{A_{v1}}{2} - \dfrac{A_{v3}}{2}$
1	80	0	20							
2	60	20	20							
3	50	30	20							
4	40	40	20							
5	30	50	20							
6	20	60	20							
7	0	80	20							

8.2.1(B). Determination of 2,4-Tolylene Diamine and 2,4-Tolylene Diurea in the Presence of Urea, with Baseline Correction Method

Principle of the Technique

The starting substance of the synthesis is 2, 4-tolylene diamine, from which the said end product is prepared by carbonylation ($COCl_2$), and subsequent ammonization. However, different side reactions also take place and, among others, a substantial quantity of urea is formed. To follow the synthesis, the concentrations of the initial substance and of the desired end product must be determined.

2,4-Tolylene diamine exhibits at the wave number 855 cm^{-1} an intensive band free of interference, which arises from the wagging vibration of the NH_2 group. A band similarly specific of 2,3-tolylene diurea occurs at a wave number of 1365 cm^{-1}, which is due to the symmetric deformation vibration of the CH_3 group.

The infrared spectrum of the reaction product is taken on the dry substance in potassium bromide pellets. Synthetic standards are used for the analysis, which are prepared by mixing the compounds to be determined, urea, and potassium bromide.

Apparatus

> IR spectrophotometer
> Agate mortar
> Table press

Chemicals

> 12 mg 2,4-tolylene dinamine; $CH_3C_6H_3(NH_2)_2$
> 12 mg 2,4-tolylene diurea; $CH_3C_6H_3(NHCONH_2)_2$
> 6 mg urea
> 10 g potassium bromide (spec. pure)

Stock Mixtures

> (A) 12 mg of 2,4-tolylene diamine + 3.00 g KBr
> (B) 12 mg of 2,4-tolylene diurea + 3.00 g KBr
> (C) 6 mg of urea + 1.50 g KBr

Note: KBr is admixed in three portions to the organic substance in the agate mortar.

Standard Mixtures

Marking	Stock mixture (g)		
	A	B	C
S1	0.80	—	0.20
S2	0.60	0.20	0.20
S3	0.50	0.30	0.20
S4	0.40	0.40	0.20
S5	0.30	0.50	0.20
S6	0.20	0.60	0.20
S7	—	0.80	0.20

Note: The weighed substances are thoroughly mixed in the agate mortar.
Column A gives also the weight ratio of 2,4-tolylene diamine, column B that
of 2,4-tolylene diurea in the organic substance.

Sample Mixtures

To 6 mg of the dried sample 1.50 g of KBr is added, and thoroughly mixed in
an agate mortar.

Preparation of the Pellets

From the mixture with KBr, 1.0 g is weighed into the die of the press, and
compressed at a pressure of 250 atm for 2 min. From a further 1.0 g of pure
KBr a pellet is prepared under the same conditions, which is used as reference
substance.

Procedure

The instrument is put into operation according to the directions for use.

Evaluation

The transmittance (T) of the peak at 855 cm^{-1} is read on the instrument, and the
average background transmittance (T_o) below the peak is determined. The
relative absorbance value is calculated from the relationship $A = lg\ (T_o/T)$. The
A values obtained for the standards are plotted as a function of the percent
concentration of 2,4-tolylene diamine, and a linear analytical curve is obtained.

The absorbance value of the characteristic band (1365 cm^{-1}) of 2,4-tolylene
diurea is calculated in a similar way, and used for the determination of the
corresponding analytical curve.

By aid of the table, the following calibration curves are prepared: absorbance calculated at 855 cm^{-1} vs. concentration of 2,4-tolylene diamine concentration and absorbance calculated at 1365 cm^{-1} vs. 2,4-tolylene diurea concentration.

Calculate the absorbance values of the unknown samples in a manner identical to that used with the calibration series at the positions 855 cm^{-1} and 1365 cm^{-1} and read the 2,4-tolylene diamine and 2,4-tolylene diurea concentration values from the calibration diagrams. The amount of urea is the missing part up to 100%.

8.2.2. Determination of Cobalt and Chromium Ions in the Presence of Each Other

Principle of the Determination _____

The molecular absorption coefficients are determined with solutions of cobalt and chromium ions at two selected wavelengths ($\lambda 1$ and $\lambda 2$) at one of which one of the components and at the other of which the other of the components absorbs predominantly. A set of equations with two unknowns can be written for the absorbance values; the unknown concentrations can be obtained by solving these equations.

$$A^{\lambda 1} = \varepsilon_{Co}^{\lambda 1} \cdot 1 \cdot c_{Co} + \varepsilon_{Cr}^{\lambda 1} \cdot 1 \cdot c_{Cr}$$

$$A^{\lambda 2} = \varepsilon_{Co}^{\lambda 2} \cdot 1 \cdot c_{Co} + \varepsilon_{Cr}^{\lambda 2} \cdot 1 \cdot c_{Cr}$$

where A are the absorbance values at positions λ_1 and λ_2, ε are the molecular absorption coefficients at positions λ_1 and λ_2, c are the unknown concentrations, and 1 is the layer thickness (the same cell is used throughout the measurement).

Apparatus

> Spectrophotometer
> Cells (1 cm layer thickness)
> Volumetric flasks, 1000 cm^3
> Volumetric flasks, 50 cm^3
> Burettes, 50 cm^3

Chemicals

> 0.2 M cobalt nitrate solution (58.210 g Co(NO$_3$)$_2 \cdot$ 6H$_2$O dissolved in 1000 cm^3 distilled water)
> 0.05 M chromium(III) nitrate solution 20.0075 g Cr(NO$_3$)$_3 \cdot$ 9H$_2$O dissolved in 1000 cm^3 distilled water

Procedure

A calibration standard series is prepared of the above stock solutions by dilution according to the following table:

Solutions Containing Co(II) Ions

Stock solution containing 0.2 M Co(II) ions (cm³)	Final volume (cm³)	Concentration (M)
10	50	0.04
20	50	0.08
30	50	0.12
40	50	0.16

Solutions Containing Cr(III) Ions

Stock solution containing 0.05 M Cr(III) ions (cm³)	Final volume (cm³)	Concentration (M)
10	50	0.01
20	50	0.02
30	50	0.03
40	50	0.04

The spectra of the 0.08 M cobalt and 0.02 M chromium calibration standard solutions are established in the 425- to 625-nm range with readings taken at every 10 nm. The absorbance-wavelength function is plotted and the peaks of the cobalt and chromium ion absorption curves are selected on the basis of the curve. These are the analytical positions (λ_1 and λ_2).

Measure the absorbance values of both standard series at the two selected wavelength values and plot these against the corresponding concentrations. The slopes of the straight lines thus obtained define the values of the molecular absorption coefficients ($\varepsilon_{Co}^{\lambda 1}$, $\varepsilon_{Cr}^{\lambda 1}$, $\varepsilon_{Co}^{\lambda 2}$, $\varepsilon_{Cr}^{\lambda 2}$) when multiplied by the layer thickness.

Hereupon the concentration measurement is carried out with the solution containing chromium and cobalt ions at an unknown concentration. For this purpose, the absorbance values of this solution are determined in the same 1-cm cell, at the selected wavelength values. In this manner, the set of equations with two unknowns, presented in "Principle of the Determination" above, and the values of $A^{\lambda 1}$ and $A^{\lambda 2}$ are obtained. By putting the molecular absorption coefficient values determined on the basis of the calibration curve into the equations, the solution of the equations gives the concentration of the sample tested.

Evaluation

Calculate the cobalt(II) and chromium(III) ion concentrations of the unknown solutions in molar units.

8.2.3. Determination of Iron in Blood Serum by Spectrophotometry

Principle of the Determination _____

The iron content of blood serum is determined after deproteination with trichloroacetic acid and reduction with hydroxyl ammonium sulfate. Iron(II) ions are reacted in a medium buffered with ammonium acetate with 4,7-diphenyl-1,10-phenanthroline disulfonic acid disodium salt (bathophenantroline disulfonate-Na), and the absorbance of the complex formed is measured at 535 nm. Synthetic standard solutions are prepared from ferrous ammonium sulfate. In healthy persons, the iron concentration of blood serum is 0.8 to 1.4 $\mu g/cm^3$; 2 cm^3 of serum is used for the test.

Apparatus

Spectrophotometer
Volumetric flask, 1000 cm^3
Volumetric flask, 500 cm^3
Volumetric flask, 200 cm^3
Volumetric flask, 100 cm^3
Pipettes 100 cm^3, 5 cm^3, 2 cm^3, 1 cm^3
Graduated pipette, 1 cm^3
Storing containers
 1000 cm^3 bottle, amber (solution 1)
 100 cm^3 polyethylene bottle (solution 2)
 200 cm^3 polyethylene bottle (solution 3)
 1000 cm^3 polyethylene bottle (solution 4)
 500 cm^3 polyethylene bottle (solution 5)
 20 cm^3 bottles, 5 pieces (standard and blank solutions)

Chemicals

Hydroxyl ammonium sulfate; $(NH_3OH)_2SO_4$
Trichloroacetic acid; CCl_3COOH
Concentrated hydrochloric acid solution
Ammonium acetate; CH_3COONH_4
4,7-Diphenyl-1,10-phenanthroline disulfonic acid disodium salt
 (bathophenanthroline disulfonate-Na; BPDS-Na)
Ferrous ammonium sulfate; $(NH_4)_2Fe(SO_4)_2 \cdot 6H_2O$
Concentrated sulfuric acid
Deionized water

Reagent and Stock Solutions

1. Protein precipitating solution: 25 g hydroxyl ammonium sulfate, 150 g trichlo-roacetic acid, and 57 cm³ of concentrated hydrochloric acid are dissolved to give a solution of 1000 cm³.
2. Buffer solutions: 100 g ammonium acetate is dissolved to give a solution of 100 cm³.
3. Reagent solution: 100 mg of BPDY-Na is dissolved in 100 cm³ of water, and 100 cm³ of solution 2 is added.
4. Stock solution (I): 2 cm³ of concentrated sulfuric acid is added to 1.404 g of ferrous ammonium sulfate and the volume of the solution is made up to 1000 cm³ (2 mg of Fe/cm³).
5. Stock solution (II): 1 cm³ of concentrated sulfuric acid is added to 5 cm³ solution 4 and the volume of the solution is made up with water to 500 cm³ (200 µg of Fe/cm³).

Note: Analytical grade reagents are used, and the solutions are prepared with deion-ized water.

Synthetic Standard and Blank Solution

Marking (µg Fe/cm³)	FeI	FeII	FeIII	FeIV	Fe$_B$
	0.167	0.333	0.500	0.667	—
Stock solution			(cm³)		
5	1	2	3	4	—
1	4	4	4	4	4
3	4	4	4	4	4
H₂O	3	2	1	—	4

Note: The solutions are shaken, and kept standing for 10 min. Photometry is carried out within 1 h.

Test Solutions

Blood serum is separated by centrifuging from the blood sample. To 2 cm³ of serum 2 cm³ of solution 1 is added, the mixture is shaken, and after 15 min centrifuged at a high speed. To 1 cm³ of the clear solution obtained 1 cm³ of solution 3 is added, the mixture is shaken, and kept standing for 10 min. Photometry is carried out within 1 hour. (The dilution of the serum is threefold.)

Procedure

The instrument is put into operation according to the directions for use.

The standard and the test solutions and the blank solution are introduced into 1-cm cells and the absorbance is measured.

Evaluation

An analytical curve is plotted from the absorbance and concentration values of the standard solutions, which is a straight line starting from the origin. The concentrations belonging to the absorbance data of the test solution are read from this line, and multiplied by three in consideration of the dilution used.

8.2.4. Determination of Small Amounts of Copper After Separation by Extraction

Principle of the Determination _____

Copper(II), similarly to several other ions, forms a purple chelate with dithizone (diphenyl thiocarbazone) in neutral or acidic solution:

The solubility of the chelate is higher by several orders of magnitude in apolar than in polar solvents. Thus copper(II) can be extracted from an aqueous solution of pH 2 with a solvent immiscible with water and containing dithizone (e.g., carbon tetrachloride, chloroform), in the form of the copper(II) dithizonate complex.

The distribution coefficient

$$K = \frac{C_{aq}}{C_{org}}$$

is very small; thus a single extraction is sufficient. As the green reagent is used in excess, the purple color of the complex cannot be observed visually.

If the wavelengths of the absorption maxima of the reagent and complex are far enough, no separation is necessary before spectrophotometric determination. This can be checked by comparing the spectra of the pure reagent and the pure complex. On this basis a wavelength can be selected where the absorbance of the complex is high and that of the reagent small. The calibration curve [absorbance vs. copper(II) concentration] plotted using data of measurements on solutions containing copper in known increasing concentrations and also the excess reagent has the greatest slope at this wavelength.

Apparatus

Spectrophotometer
Separating funnels with stand
Test tubes with holder
Pipette, 10 cm^3
Automatic burette, 50 cm^3
2 burettes, 20 cm^3

Chemicals

Dithizone in carbon tetrachloride
Cu(II) stock solution (1 mg Cu^{2+}/cm^3)
Deionized water

Procedure

Fill 20-cm^3 portions of the dithizone solution in carbon tetrachloride into six separating funnels using the burette. Then add 2-, 4-, 6-, 8-, 10-, and 12-cm^3 portions, respectively, of the copper(II) stock solution and deionized water to complete the volume of the aqueous phase to 12 cm^3. Close the funnels and shake them vigorously, opening the stopcock in the upside-down position of the funnel from time to time for equilibrating the pressure.

Fill 20 cm^3 of the dithizone solution into a clean separating funnel and add copper(II) stock solution until the color of the complex appears, shaking the funnel from time to time. Then place the funnels on the holder and allow the phases to separate. Remove the stoppers from the funnels, discard the first 1 to 2 cm^3 of the organic solution, and let the remainder into dry test tubes of equal diameters. Take care that the aqueous phase does not get into the tubes. Then wash the separating funnels. Pour some pure dithizone solution into a test tube.

Establish the absorbance vs. wavelength curve of the pure dithizone reagent and of the solution of the complex without excess reagent between 440 and 660 nm, increasing the wavelength in 20-nm steps. It is advantageous to make the calibration measurements at the wavelength belonging to the maximum absorbance of the complex or even more advantageous to measure at the minimum absorbance of the reagent. Measure the absorbance of the six calibration solutions at this wavelength taking two parallel measurements at each concentration.

Treat the unknown copper(II) solution similarly to the standards and measure the absorbance of the organic phase at the wavelength selected.

Evaluation

Plot the calibration curve and give the concentration of the unknown in μg/100 cm^3.

8.2.5. Determination of Manganese in the Form of Permanganate

Principle of the Determination _____

Manganese(II) is oxidized with potassium peroxydisulfate in the presence of a small amount of silver nitrate (used to remove chloride possibly present) and phosphoric and nitric acids at boiling temperature. Phosphoric acid is necessary to prevent precipitation of manganese dioxide and to ensure oxidation to permanganate. The concentration of silver may be chosen low so that no silver chloride precipitation is observed if chloride is only present in a small amount. The concentration of nitric acid should not exceed 0.3 *M*. Interference from chloride may be eliminated by adding some mercuric sulfate which forms a slightly dissociated complex with chloride. Bromide and iodide are tolerated if present at a very low concentration.

Manganese can be determined in the presence of some organic matter by heating for a longer period of time and adding more peroxydisulfate.

The permanganate solution formed is stable for at least 24 hours at room temperature if peroxydisulfate is present in excess and no reducing agent is present.

Apparatus

Spectrophotometer
Volumetric flasks, 100 cm^3, 1000 cm^3
Graduated cylinders, 10 cm^3, 100 cm^3, 1000 cm^3
Conical flask, 200 cm^3
Precision balance
Bunsen burner
Asbestos wire gauze

Chemicals

Mercuric sulfate
Concentrated nitric acid
Concentrated phosphoric acid
Silver nitrate
Ammonium or potassium peroxydisulfate
Special reagent: dissolve 75 g mercuric sulfate in 400 cm^3
concentrated nitric acid and 200 cm^3 of water; add 300 cm^3 of
85% phosphoric acid and 0.035 g silver nitrate; after cooling,
make up the solution to 1000 cm^3

Procedure

Add 5 cm^3 of the special reagent to the sample solution which may contain 0.1
g sodium chloride as a maximum. Dilute the solution to 90 cm^3 with distilled
water, add 1 g peroxydisulfate and heat to boiling. Boil in a 200 cm^3 conical flask,
and add a glass bead against retarded boiling. Boil the mixture for 2 min, then
allow to stand for 1 min and cool under tap water. Transfer the solution to a
100-cm^3 volumetric flask and make up to the mark with distilled water which
may not contain any reducing matter. (Purify the water previously by distillation
in the presence of potassium permanganate.) Measure the absorbance of the
solution in a 2-cm cell at 525 nm. Determine the concentration using a calibration
curve, prepared by plotting the absorbance of solutions prepared by treating
solutions of known Mn(II) concentrations in the same way as the sample.

If the amount of manganese in the sample taken exceeds 2 mg, some
manganese dioxide may precipitate. In such cases add 2 cm^3 of 85% phosphoric acid to the solution, reduce the manganese dioxide with one or two drops
of 10% hydroxylamine solution, and boil again to transform manganese(II) to
permanganate completely.

If some organic matter is present, boil for a longer period of time (5 min
or more) and use more peroxydisulfate (2 to 5 g).

Evaluation

Plot the calibration curve and calculate the Mn(II) concentration in the unknown solution in μg/cm^3.

8.2.6.   Determination of Phosphorus

Principle of the Determination _____

Phosphorus is quantitatively isolated from all interferences prior to spectrophotometric determination. Extraction according to Erdey and Fleps with i-butyl
alcohol can be used to advantage, as the solvent extracts the phosphomolybdate
heteropolyacid selectively. The essence of the determination is as follows:
ammonium phosphomolybdate complex prepared in acidic medium is extracted into i-butyl alcohol and, after washing the organic phase with sulfuric
acid, the complex is reduced with an acidic solution of tin(II) chloride to
molybdenum blue and the latter is measured by spectrophotometry. Care
should be taken that the extraction is complete and also that water traces
containing ammonium molybdate are completely removed from the i-butyl
alcoholic phase by washing with acid, since any ammonium molybdate present
is also reduced to molybdenum blue thus giving rise to a positive error.

Beer's law is valid between 0.1 and 1.5 $\mu g/cm^3$ phosphorus content.

Apparatus

> Spectrophotometer
> Burette, 12 cm^3
> Pipette with two marks, 10 cm^3
> Pipette, 10 cm^3, graduated
> Separating funnel
> Volumetric flasks, 25 cm^3

Chemicals

> Potassium dihydrogen phosphate solution 0.1 to 1.5 $\mu g/cm^3$;
>   10 $\mu g/cm^3$
> Tin(II) chloride solution: dissolve 2 g $SnCl_2 \cdot 2H_2O$ in
>   10 cm^3 1:1 hydrochloric acid with warming; dilute
>   2 cm^3 of this solution with 98 cm^3 of 1 N H_2SO_4
> 1 N sulfuric acid
> i-Butyl alcohol
> Ethyl alcohol
> 5% ammonium molybdate solution

Procedure

Bring the unknown solution to the separating funnel and fill up to about 10 cm^3.
Add 2.5 cm^3 of 5% ammonium molybdate solution and 2 cm^3 of 1 N H_2SO_4.
Add 10 cm^3 i-butanol and shake vigorously for 2 min. After the phases have
separated, discard the aqueous phase and wash the organic phase with three
5-cm^3 portions of 1 N H_2SO_4, shaking for half a minute each time. After

removing the acid, add 10 cm³ of freshly diluted tin(II) chloride solution to the organic phase, shake for half a minute, allow the phases to separate, discard the aqueous phase, and transfer the organic phase to a 25 cm³ volumetric flask using ethyl alcohol for rinsing and for making up the solution to the mark. Wait for 10 min and measure the absorbance in a 1-cm cell at 527 nm, against a blank prepared in the same way as the sample but using distilled water instead of the phosphorus-containing sample.

Any blue coloration of the blank is indicative of the contamination of either of the reagents or of incomplete removal of the excess of ammonium molybdate reagent. Tin(II) chloride has to be freshly diluted.

To calibration measurements prepare solutions of concentrations to obtain 0.20, 0.30, 0.40, 0.50, 0.60, 0.80, and 1.5 µg P/cm³ in the final, organic solution. In preparing the diluted series, consider that the final volume is 25 cm³. Treat the calibration solutions similarly to the sample and measure the absorbance of the solution obtained.

Evaluation

Plot the calibration curve and calculate the phosphorus contained in the sample in µg/cm³.

8.2.7. Determination of Iron(III) with Sulfosalicylic Acid

Principle of the Determination

Iron(III) forms a complex with sulfosalicylic acid, the absorbance of which can serve as the basis for spectrophotometric determination. To prevent hydrolysis of the complex, the measurement has to be carried out at pH 1.

Apparatus

Spectrophotometer
Burette, 12 cm³
Pipette, 15 cm³, graduated
Beakers, 50 cm³
Beakers, 300 cm³
Volumetric flasks, 50 cm³

Chemicals

0.01 M iron(III) solution in 0.1 M HCl₄
0.01 M sulfosalicylic acid in 0.1 M HClO₄
0.1 M perchloric acid

Procedure

Rinse the sample solution into a 50-cm³ volumetric flask with 0.1 M perchloric acid, add 30 cm³ of 0.01 M sulfosalicylic acid, and make up to the mark with 0.1 M perchloric acid. Allow to stand for half an hour and measure the absorbance in a 1-cm cell at 500 nm.

Determine the concentration of the unknown using a calibration graph. To prepare the calibrating solutions, pour 1.00, 2.00, 3.00, 4.00, and 5.00 cm³ of the 0.01 M iron(III) solution into 50-cm³ volumetric flasks, treat as the unknown, and after half an hour's standing measure the absorbances.

Evaluation

Plot the calibration curve (absorbance vs. concentration, $\mu g/cm^3$) and determine the iron(III) concentration of the unknown in $\mu g/cm^3$.

8.2.8. Determination of the Concentration of Methyl Red at the Isobestic Point

Principle of the Determination _____

Methyl red is a reversible acid-base indicator. The color and absorbance of its solution depend on pH. The concentration of the indicator is best measured at the isobestic point.

Apparatus

> Spectrophotometer
> Volumetric flasks, 100 cm³
> Pipettes 5 cm³, 25 cm³, 50 cm³, all with two marks
> Burette, 12 cm³
> Beakers, 250 cm³

Chemicals

> $10^{-3}\%$ methyl red solution
> 0.04 M sodium acetate
> 0.02 M acetic acid

Procedure

Determination of the Isobestic Point of the Indicator

Prepare indicator solutions of pH 4.73 and 5.73 as follows. Pour solutions according to the following table into two 100-cm³ volumetric flasks

	pH = 4.73	pH = 5.73
10^{-3}% indicator solution	10.00 cm³	10.00 cm³
0.04 M sodium acetate	25.00 cm³	25.00 cm³
0.02 M acetic acid	50.00 cm³	5.00 cm³

Fill up the solutions to the mark with distilled water. Then take the spectra of both solutions by measuring the absorbances from 450 to 470 nm, changing the wavelength in 2-nm steps. Plot the two spectra in one diagram. Determine the wavelength belonging to the isobestic point, i.e., the intersection point of the two spectra.

Determination of the Concentration of Methyl Red

Fill up the unknown solution to 100 cm³ in a volumetric flask and measure its absorbance at the isobestic point. The concentration to be determined is, on the basis of Beer's law:

$$c = \frac{E}{E_{1\%}^{1\,cm} \cdot 1}$$

where $E_{1\%}^{1\,cm} = 531$; E is the absorbance measured; and 1 is the optical path length, cell length in centimeters.

Evaluation

Calculate the concentration of methyl red in $\mu g/cm^3$.

8.2.9. Determination of the Active Ingredient (5-Ethyl-2′-Deoxyuridine) of the Ointment *Revidur*

Principle of the Technique _____

UV spectrophotometric and NIR spectroscopic methods were developed for determining the active ingredient (5-ethyl-2′-deoxyuridine) in the antiviral ointment (Revidur).

UV spectrophotometric measurements were carried out at 263 nm with solution of the ointment in absolute ethanol prepared by ultrasonication. Calibration solutions contained the matrix. The standard error of the method was 0.016%; the correlation coefficient of the calibration curve was 0.999 in the range 0.25 to 2.5% drug content.

Two NIR methods are described in the present work. In one of the methods, NIR reflectance spectra were taken of solutions and concentration determinations were carried out based on the second derivative of the spectra at 1166 nm.

The correlation coefficient was equal to that calculated for the UV method, and the standard calibration error was slightly higher, 0.05%. The advantage of the method is its quickness if calibration data are available.

In the present work, a method has been worked out for the determination of the active principle in the ointment Revidur which involves no extraction step and applies calibration measurements in the presence of the ointment matrix to eliminate matrix effects. The composition of the ointment is as follows (for 10 g total weight):

> Drug 0.1 g
> Sorboxaethenum stearinicum 0.40 g
> Paraffinum liquidum 0.40 g
> Alcohol cetylstearylicus 1.20 g
> Vaselinum album 2.00 g
> Solution conservans 0.20 g
> Aquadestillata 5.70 g

8.2.10. Determination of the Drug by UV Spectrophotometry

Apparatus and Chemicals

The drug content was determined in absolute ethanol solution. By sonication all the components of the ointment were dissolved in absolute ethanol which could not have been achieved otherwise. A series of calibration solutions was prepared by mixing the solution of the matrix and that of the drug in appropriate proportions.

Ointment samples were dissolved in absolute ethanol by sonication.

Absorbance measurement was carried out by a UV-VIS spectrophotometer.

- The solution of the matrix: 0.1 g matrix was weighed on an analytical balance, transferred to a 100-cm^3 volumetric flask with 80 cm^3 absolute ethanol and sonicated for 4 min. The solution thus obtained was filled up to the mark with absolute ethanol.

- Sample solution was prepared similarly, using ointment portions depending on the expected drug concentration (0.1 g in the case of 1% and 0.08 g in the case of 2.5% drug content).

- Drug stock solution: 0.05 g of the drug was dissolved in absolute ethanol similarly to the above solutions.

Calibration solutions were prepared as given in Table 1.

Table 8.1
Data of the Series of Calibration Solutions for the Determination of the Drug in the Ointment Revidur by UV Spectrophotometry

Drug stock solution (cm³)	0.05	0.10	0.20	0.30	0.50
Solution of the matrix (cm³)	9.95	9.90	9.80	9.70	9.50
Drug concentration (in % of the matrix)	0.25	0.50	1.00	1.50	2.50

Procedure and Evaluation

The UV spectra of the solution of the matrix and of the two ointment samples were taken in the range of 200 to 300 nm and it was found that the matrix components did not interfere in the determination of the drug if the solution of the matrix was used as reference solution (Figure 8.6).

The spectra of the calibration series are shown in Figure 8.7.

Five parallel series were prepared, each series consisting of five members.

The parameters of the equation describing the concentration dependence of the absorbance measured at 263 nm, the correlation coefficient, and the standard error of the calibration (for interpretation see the next chapter) were calculated:

$$A_{263\,nm} = -0.008 + 0.351c \ [\%]$$

$$\tau = 0.999$$

$$\text{standard calibration error} = 0.016 \ [\%]$$

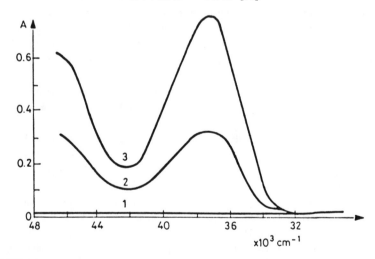

FIGURE 8.6

UV spectra of the solutions of the ointment matrix and of two Revidur ointment samples in absolute ethanol. (1) Matrix; (2) Revidur sample with 1% drug content; (3) Revidur sample with 2.5% drug content.

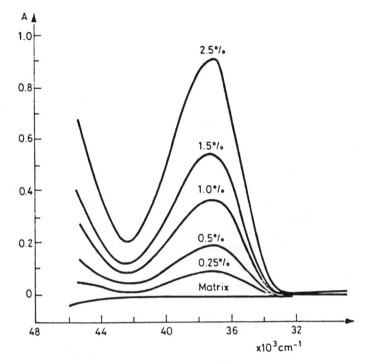

FIGURE 8.7
UV spectra of the calibration solutions.

The above equation was used for determining the drug in various preparations and the results were always close to the values expected.

The recovery of added drug was also studied. Some of the results are as follows:

Theoretical concentration (%)	1.60	2.10	2.59
Measured concentration (%)	1.58	2.08	2.56

8.2.11. Determination of the Drug by NIR Spectroscopy

Apparatus and Chemicals _____

Calibration solutions were prepared as described previously. The drug content of the series members in percent of the matrix were 0.25, 0.50, 1.0, 1.5, and 2.5. In preliminary experiments, NIR spectra of the calibration solutions were taken against the internal reference of the instrument. The latter gave better results.

The measurements were made using a NIR instrument.

Procedure and Evaluation

Three spectra were taken of each calibration solution and the computer program of the instrument processed the second derivative of the averaged spectrum to find the wavelength at which there is correlation between the spectrum or its second derivative and the concentration of the drug and to determine the parameters of the calibration curve, correlation coefficients, and other statistical parameters.

The reflectance spectrum and its second derivative for a calibration solution are shown in Figure 8.8; the calibration curve calculated for the second derivative at 1166 nm, with the standard deviation of the deviation of concentrations determined based on the calibration curve from actual concentrations, is shown in Figure 8.9.

The value of the correlation coefficient *(r)* reflects a close correlation between optical and chemical data.

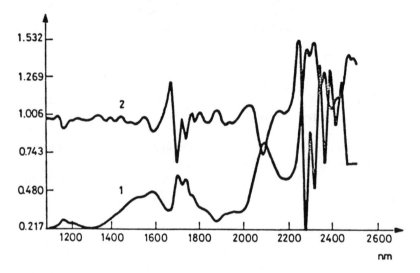

FIGURE 8.8

NIR reflectance spectrum (1) and its second derivative (2) taken of a calibration standard solution.

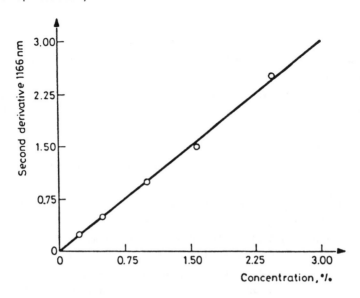

FIGURE 8.9
Calibration curve at 1166 nm. Standard error of calibration: 0.05%, $r = 0.999$.

8.3. References

Bellamy, L., *The Infrared Spectra of Complex Molecules,* Methuen and Co., London, 1962.

Cleve, G., Hoyer, G. A., Schulz, G., and Verbruggew, H., *Chem. Ber.,* 106, 3062, 1973.

Cross, A. D., *An Introduction to Practical Infrared Spectroscopy,* Butterworths, London, 1960.

Csarnyi, A. H., Szabolcs, A., Vajda, M., and Ötvös, L., *J. Chromatogr.,* 169, 426, 1979.

Graze, G., Alexander, G., and Till, A., *J. Chromatogr.,* 191, 253, 1980.

Hempel, B., *Deutsch. Apoth. Ztg.,* 122, 1670, 1982.

Kaul, R. and Hempel, B., *Arzneim. Forsch.,* 35, 1066, 1985.

Kemin, M., Eikel, I., and Shugar, D., *Eur. J. Biochem.,* 53, 197, 1975.

Kiss-Eröss, K., Vakulya, G., Fábián, Z., and Balogh, S., *Per. Polytechn. Chem. Eng.,* 35, 197, 1991.

Kiss-Eröss, K., Analytical infrared spectroscopy, in *Wilson and Wilson's: Comprehensive Analytical Chemistry,* Vol. VI, Svehla, G., Ed., Elsevier, Amsterdam, 1976.

Mellon, M. G., *Analytical Absorption Spectroscopy,* John Wiley & Sons, New York, 1950.

IR

Bellamy, L. J., *The Infrared Spectra of Complex Compounds,* Chapman and Hall, New York, 1980.

Jones, N. and Sandorfy, C., Infrared and raman spectrometry, in *Chemical Applications of Spectroscopy,* Vol. 9, Weissberger, A., Ed., Interscience, New York, 1956.

Nyquist, R. A. and Kagal, R. O., *Infrared Spectra of Inorganic Compounds,* Academic Press, New York, 1971.

Socrates, G., *Infrared Characteristic Group Frequencies,* John Wiley & Sons, Chichester, England, 1980.

Clerc-Pretsch-Seibl-Simon, *Tabellen zur Strukturaufklärung organischer Verbindungen mit spektroskopischen Methoden,* Springer-Verlag., Heildelberg, 1981.

Colthup, N. B., Daly, L. H., and Wiberley, S. E., *Introduction to Infrared and Raman Spectroscopy,* Academic Press, San Diego, 1990.

Ferraro, J. R. and Krishnan, K., Eds., *Practical Fourier Transform Infrared Spectroscopy, Industrial and Laboratory Chemical Analysis,* Academic Press, San Diego, 1990.

Collections of Spectra

Aldrich Library of Infrared Spectra, Aldrich Chem. Co., Milwaukee, 1975.

Sadtler, S. P., Infrared Standard Spectra, Sadtler Research Laboratories, Philadelphia, supplemented continuously.

UV-VIS

Burgess, C. and Knowles, A., Techniques in Visible and Ultraviolet Spectrometry, in *Standards in Absorption Spectrometry,* Chapman and Hall, London, 1981.

Koch-Koch-Dedic, *Handbuch der Spurenanalyse,* Springer-Verlag, Berlin, 1974.

Kortüm, G., *Kolorimetrie, Photometry und Spektroskopie,* Springer-Verlag, Berlin, 1955.

Marczenko, Z., *Spectrophotometric Determination of Elements,* Harwood-Wiley, Chichester, England, 1976.

Rao, C. N. R., Ultraviolet and visible spectroscopy, in *Chemical Application,* 3rd ed., Butterworths, London, 1975.

Collections of Spectra

Hirayama, K., *Handbook of Ultraviolet and Visible Absorption Spectra of Organic Compounds,* Plenum Press, New York, 1967.

Láng, L., *Absorption Spectra in the Ultraviolet and Visible Region,* Academic Press, New York, 1959, 1.

Chapter 9

Fluorescence Analysis (Fluorimetry)

Principle of the Technique

Fluorescence occurs if an atom or molecule is excited by a photon and it reemits part of its energy in the form of a radiation, while it returns to the ground state. If the duration of the excited state is about 10^{-7} s, the phenomenon is termed fluorescence. If a longer period of time (10 to 10^{-4} s) elapses between excitation and emission, we speak of phosphorescence.

Analytical procedures based on the phenomenon of fluorescence are termed fluorimetry. The great majority of compounds which fluoresce with such an intensity as to be applicable for analytical purposes belongs to the aromatic compounds. The condition of the fluorimetric determination of metal ions is that it should be possible to convert the metal with an appropriate complex-forming agent into a chelate complex capable to show fluorescence.

The wavelength range of the absorbed radiation may range from visible light to X-rays. Absorption of visible and ultraviolet light brings about changes in the electron state of the molecules. When plotting the intensity of the fluorescent light emitted by the excited molecule against the wavelength of the exciting light, a fluorescence spectrum — characteristic of the quality of the molecule — is obtained. The intensity of fluorescent light is, in the proper concentration range (generally up to 10 $\mu g/cm^3$) proportional to the concentration of the substance emitting fluorescent light.

9.1. Observations on Fluorescence by Aid of a Fluorescein Solution

Principle of the Determination _____

The intensity of fluorescent light is dependent on the experimental conditions, such as the wavelength of the exciting light, the concentration of the fluorescent substance, and the pH. The following measurements serve to study the influence of these parameters.

Apparatus

> Spectrophotometer Zeiss "Spekol"
> Fluorimetric adapter with cells and amplifier
> Volumetric flask, 1000 cm^3
> Volumetric flask, 100 cm^3
> Burette, 50 cm^3
> Graduated cylinder

Chemicals

> Fluorescein solution, 100 $\mu g/cm^3$
> Acid mixture used in Britton-Robinson buffer solution
> Sodium hydroxide solutions, 1 N 0.2 N

(a) Establishment of the excitation spectrum of fluorescein with application of a mercury vapor lamp.

Principle of the Determination _____

The spectrum is made with a solution of 80 $\mu g/cm^3$ concentration, prepared from a solution of 100 $\mu g/cm^3$ by dilution with 1 N sodium hydroxide solution.

The amplification on the instrument is adjusted so as to enable an adjustment of 50% deflection on the scale of the meter built into the Spekol instrument at 440 nm by means of the knob marked 100 on the amplifier.

The wavelength of the exciting light is increased between 340 and 500 nm is steps of 10 nm or — where the intensity of the fluorescent light changes to a great extent — in steps of 2 nm. The deflection of the meter is read at each wavelength on the scale marked T.

Evaluation

Construct the excitation spectrum on the basis of the measured data and select the wavelength corresponding to the excitation of the highest intensity. The further measurements — described under points (b) and (c) — are carried out at this wavelength.

(b) Examination of the intensity fluorescent light of fluorescein as the function of pH.

Procedure

Britton-Robinson buffer mixtures are prepared according to the following table:

Britton-Robinson fundamental acid mixture (cm³)	0.2 N NaOH (cm³)	pH
100	0	1.8
100	25	4.1
100	35	5.0
100	38.5	5.5
100	42.5	6.1
100	47.5	6.6
100	52.5	7.0
100	60.0	8.0

Prepare a series of solutions from the fluorescein solution of 100 µg/cm³ concentration with each of the buffer solutions in such a manner that a 25-cm³ portion of the fluorescein stock solution is filled up to the mark in a 100-cm³ volumetric flask with the buffer solution of the given pH value.

10 cm³ 1 N NaOH solution and 25 cm³ of the 100 µg/cm³ fluorescein solution are also filled up to 100 cm³ with distilled water in a volumetric flask. The latter solution is used to adjust 100% on the scale marked T; the intensity of fluorescent light is measured with the series of solutions prepared according to the above in the order of increasing pH values.

Evaluation

The intensity of fluorescent light is plotted against pH.

(c) Examination of the connection between intensity of fluorescent light and concentration.

Procedure

The following series of solutions is prepared from a stock solution containing 100 µg/cm³ fluorescein (stock solution I) by dilution with distilled water:

Stock solution I (cm³)	1 N NaOH (cm³)	Final volume (cm³)	Concentration (μg/cm³)
1	10	100	1
4	10	100	4
7	10	100	7
10	10	100	10
15	10	100	15
20	10	100	20
25	10	100	25
30	10	100	30
40	10	100	40
50	10	100	50
65	10	100	65
80	10	100	80

A light intensity of 80% is adjusted with the solution of 80 μg/cm³ concentration. The fluorescence of the solutions is measured in the order of increasing concentrations in such a manner that prior to each reading the 80-μg/cm³ solution is placed into the cell holder, and the reading of 80% is corrected, should there be any deviation from it.

The meter deflection readings pertaining to the different concentration values, as read on the scale marked T, are noted.

Hereupon a 10-cm³ portion of the 100-μg/cm³ stock solution is diluted to 1000 cm³ with distilled water. From the solution of 1-μg/cm³ concentration thus prepared (stock solution II) the following are prepared by dilution with distilled water:

Stock solution II (cm³)	1 N NaOH (cm³)	Final volume (cm³)	Concentration (μg/cm³)
5	10	100	0.05
10	10	100	0.10
15	10	100	0.15
20	10	100	0.20
25	10	100	0.25
30	10	100	0.30
35	10	100	0.35
40	10	100	0.40
45	10	100	0.45
50	10	100	0.50

A deflection of 100% is adjusted with the 0.5-µg/cm³ solution and the intensities of the fluorescent light of the solutions are measured as in the above.

Evaluation

Plot the concentration (C) vs. meter deflection (T) diagram from the results of both measurement series.

9.2. Fluorimetric Determination of α-Tocopherol

Principle of the Determination _____

The name tocopherol or vitamin E relates to a group of compounds of vitaminlike action. All these compounds are derivatives, methylated to various degrees and at various positions on the aromatic ring of the basic compound tocol which is 2-methyl 2-(4,8,12 trimethyl tridecyl) 6-hydroxy chromane:

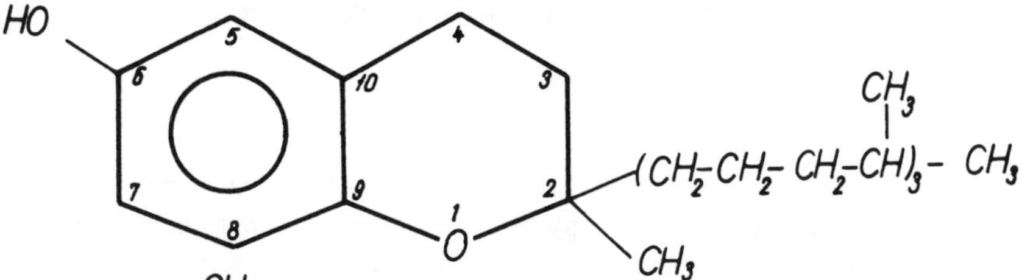

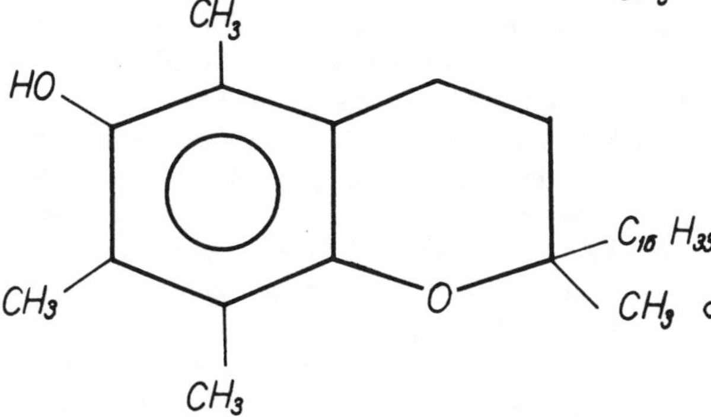

The techniques used for the quantitative determination of vitamin E are based on the fact that α-tocopherol is easy to oxidize. In pharmaceutical preparations and, partly, also in natural form, it is found in the form of its esters (e.g., acetate) which have to be hydrolyzed prior to determination. α-Tocopherol is highly sensitive to light and oxygen and consequently the preparation of samples has to be carried out in an atmosphere of nitrogen and with the exclusion of light.

In the fluorimetric determination of α-tocopherol, the compound is oxidized to tocopherol red and thereupon the compound of o-quinoidal structure is made to condense with o-phenylene diamine. The phenazine product obtained possesses fluorescent activity and can be separated from the side products by chromatography on an aluminum oxide column.

The condensation reaction goes on at the highest rate and quantitatively in the case of α-tocopherol quinone. β-Tocopherol quinone reacts at a lower rate and the fluorescent intensity of its phenazine derivative is about half of that of α-tocopherol. The reaction of γ- and δ-tocopherol is even slower.

Apparatus

Spectrophotometer Zeiss "Spekol"
Fluorimetric adapter with cells and amplifier

Chemicals

α-Tocopherol phenazine solution, 1 to 8 $\mu g/cm^3$ in a 1:9 mixture of butyl alcohol and ethyl alcohol
Solvent mixture butyl alcohol-ethyl alcohol, 1:9

Procedure

a. Establishment of the excitation spectrum of the phenazine derivative of α-tocopherol with application of a mercury vapor lamp. The spectrum is made with an 8-$\mu g/cm^3$ solution. The deflection of the meter is adjusted to 80 to 90% on the T scale. The exciting wavelength is varied in the 320 to 470-nm range and the deflection of the meter is read at the different wavelengths on the scale marked T.

b. Establishment of a calibration curve. The fluorescent light intensities of α-tocopherol samples of the concentrations 1, 2, 3, 4, 5, 6, and 8 $\mu g/cm^3$ are measured at a wavelength of 365 nm. The 0% value is adjusted with the pure solvent, the 100% value with the 8-$\mu g/cm^3$ sample. The measured intensity values are plotted against concentration.

c. Analysis of an unknown α-tocopherol sample. The fluorescent light intensity of the samples of unknown concentration — marked with a serial number — is measured under conditions identical to those used when preparing the calibration curve.

Evaluation

Find the α-tocopherol content of the samples on the basis of a calibration curve established according to point (b). Plot the excitation spectrum of tocopherol.

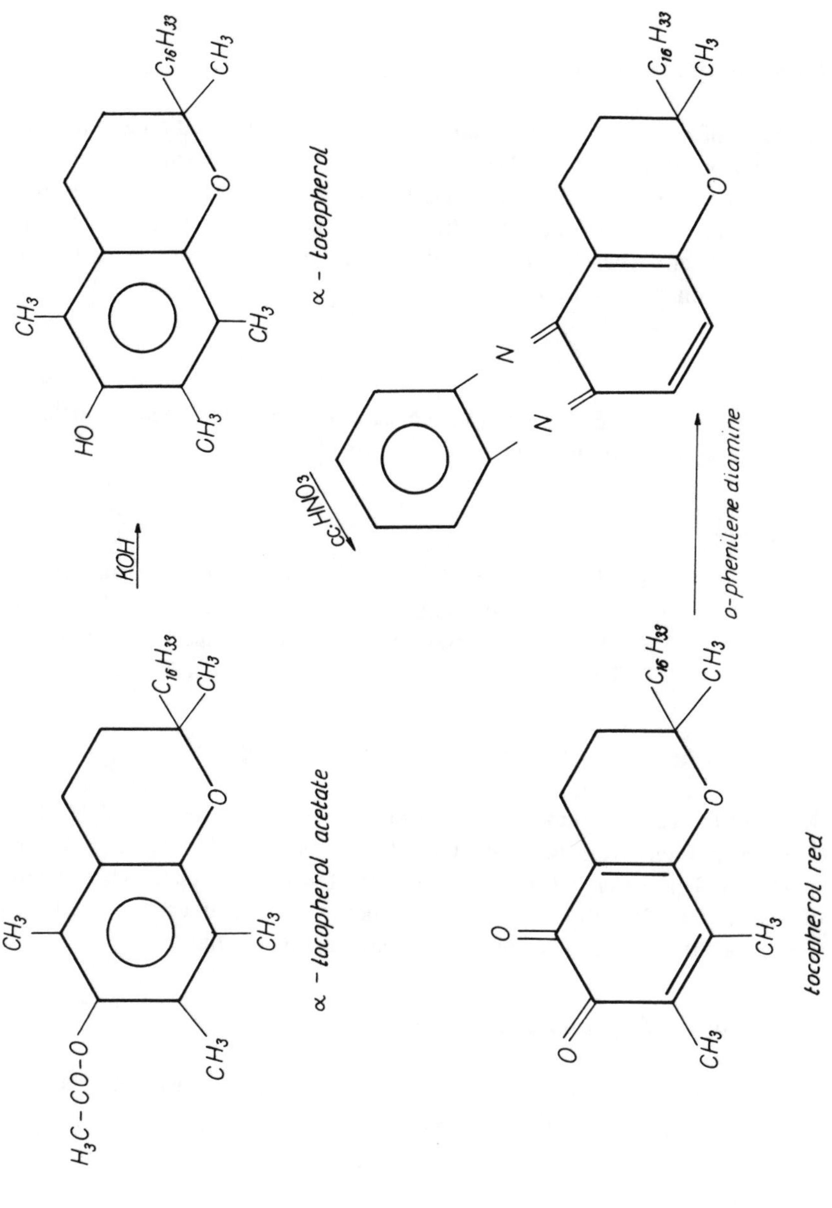

α - tocopherol

α - tocopherol acetate

tocopherol red

KOH

cc. HNO₃

o-phenilene diamine

9.3. Determination of Quinine in Tonic Water

The quinine concentration in beverages is typically 25 to 60 mg/l.

Principle of the Determination _____

Solutions of quinine fluoresce strongly when excited by radiation at 350 nm. The relative intensity of the fluorescent peak at 450 nm provides a sensitive method for the determination of quinine in beverages. Preliminary measurements are needed to define a concentration range in which fluorescent intensity is either linear or nearly so. The unknown is then diluted as necessary to produce readings within this range.

Apparatus

Fluorimeter with two monochromators or with one monochromator and one appropriate filter

Reagents

Sulfuric acid, 0.05 M and 1.00 M (for diluting the samples)
Quinine sulfate standard; weigh 0.100 g of quinine sulfate into a 1-1 volumetric flask and dilute to the mark with 0.05 M H_2SO_4; the concentration of the solution is 100 mg/l

Determination of a Suitable Concentration Range

Measure the relative fluorescent intensity of the stock solution at 450 nm. Prepare dilutions from this solution with 0.05 M sulfuric acid. Repeat the dilution and measurement process until the relative intensity approaches that of the blank (0.05 M sulfuric acid). Make a plot of the data and select a suitable range for the analysis, a region within which the plot is linear.

Preparation of a Calibration Curve

Prepare four or five standards that span the linear region; measure the fluorescence intensity for each. Plot the data.

Procedure

Make suitable dilutions from the unknown containing 0.05 M H_2SO_4 to bring its fluorescence intensity within the calibration range.

Evaluation

Calculate the quinine concentration of the unknown in mg/l.

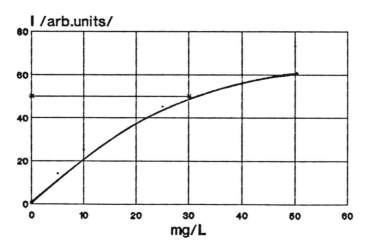

FIGURE 9.1
Typical calibration curve for quinine.

Chapter 10

Atomic Spectroscopy

10.1. Atomic Absorption Spectroscopy

Principle of the Technique

Atomic absorption spectrometry (AAS) is perhaps the most widely used method for metal and trace metal analysis. The AAS method is characterized by excellent selectivity, good detection limits, precision of 0.5 to 2% under optimal conditions, and relative freedom from interferences. Atomic absorption is a special technique of UV-VIS spectrometry covering the wavelength range from 190 to 860 nm based on the excitation of free atoms in the ground state by photons. Free atoms are produced in atomizers of different constructions, such as flame atomizer or electrothermal atomizer, in both cases at high temperature from 1000 to 3000 K. In the vapor of metallic mercury there are free atoms present even at ambient temperature allowing the cold vapor atomic absorption determination of this element.

Atomic absorption corresponds to electron transitions from low to higher energy states of free atoms in an excitation process by energy carried by photons. The wavelength of the absorption lines is defined very accurately by the atomic energy levels of electrons, the typical half width of lines being around 0.005 nm. The overlap of the absorption lines of different elements is almost negligible; therefore, the selectivity of the method is outstanding.

The measurement of atomic absorption needs a monochromatic radiation, which means a band width of 0.001 nm or less and the wavelength should be set with a similar accuracy.

To avoid the use of very expensive high-resolution monochromators Walsh (1953) suggested the use of atomic spectral line source of the element measured as light source combined with a medium resolution monochromator with spectral band width of 0.2 to 1 nm. The commonly used atomic absorption light sources are the hollow cathode lamps (HCL) and electrodeless discharge lamps (EDL). In both lamps a kind of vacuum discharge with the excited free atoms of the analyte present in it is the source of emission which guarantees the coincidence of line wavelengths only for the element measured, known as the "lock and key" effect. In this combination the monochromator has the function of the selection of a given line from the complex line spectra of the light source.

The other way for the measurement of the narrow line absorption of free atoms is with the use of intense UV-VIS continuum light source (xenon arc, IMAC) and a very high resolution monochromator with a band width 0.001 nm. Such resolution is offered by the echelle grating spectrometers; however, this kind of instrument is not commercialized. The commercially available instruments are all constructed with line source.

Atomic absorption is quantified by an equation similar to the Lambert-Beer's low in molecular spectroscopy:

$$A = \log \frac{I_0}{I} = K_\lambda \, l \, c_a \tag{1}$$

where A is the absorbance, I_0 the intensity of incident light, I that of the transmitted light, l the absorption path length, K_λ the absorption coefficient, and c_a the concentration of the free atoms of the element measured at λ wavelength.

In an analytical application of the atomic absorption principle the analyte concentration of a solution entering the atomizer c_s has to be determined. The free atom concentration in the atomizer c_a is proportional to c_s; the analyte concentration of the solution is

$$K = \frac{c_a}{c_s} \tag{2}$$

where K is a constant for the atomizer.

The analytical calibration function of AAS is

$$A = \log \frac{I_0}{I} = K \, K_\lambda \, l \, c_s \tag{3}$$

which describes a linear relationship between the absorbance and analyte concentration of the solution. In practice many calibration graphs are nonlinear, especially at high absorbance levels. One of the major sources of the curvature is the stray light resulting from the insufficient separation of the analyte line from other lines from the light source. The best precision can be reached in the range of 0.1 to 0.8 absorbance. Precision becomes significantly poorer when the slope of the calibration function has become one third that in the linear range. This point is usually used as the higher limit of the analytical range. AAS techniques, both flame and furnace, have a dynamic range of about 2.5 orders of magnitude calculated from the detection limit.

In the atomizers under the conditions of real sample analysis there may be present not only free atoms but thermally stable molecules (NaCl, KCl, etc.) and solid particles as well. The absorption by molecular species and light scattering on solid particles originated from the matrix called background absorption occurs over a broad wavelength range. The background absorption is superimposed on the atomic absorption causing a positive error. There are three background correction techniques mostly used and incorporated into AA spectrometers: (1) continuum source (deuterium arc, D_2-HCL, tungsten lamp) method, (2) method based on the use of Zeeman effect, and (3) the Smith-Hieftje method using the effect of line broadening.

Flame Atomic Absorption Spectroscopy (F-AAS)

The atomizer in F-AAS is a premixed, stationary laminar flame formed on a 5 to 10 cm long slot burner head. The burner head is connected to the nebulizer spray-chamber unit of the instrument. The sample solution is introduced into the flame in the form of wet aerosol formed in a concentric pneumatic nebulizer using the oxidant gas flow to nebulize the sample. The critical parts of the nebulizer are fabricated from corrosion-resistant materials (e.g., Ta, Pt-Ir alloy, ceramics, PTFE). The nebulizer is sucking the sample through a 20 cm long polyethylene or PTFE capillary tube with an uptake rate (q_n) of 2 to 4 ml/min and forms a primary aerosol beam. In the primary aerosol leaving the nebulizer the distribution of droplet size is wide; the average diameter is around 15 μm. In the spray chamber the sample aerosol, the oxidant gas flow (Q_{ox}), and fuel gas flow (Q_f) are mixed and the droplets above the size of 5 μm are removed by impaction and settling. The condensed solution is taken away from the spray chamber through a liquid trap.

The mixture of oxidant and fuel with the fraction of small sample droplets in it passes through the burner head and forms the flame. The efficiency of nebulization (η) around 0.06 to 0.10 is defined as the ratio of sample reaching the flame related to the one entering the nebulizer.

The most popular flames are acetylene-air (T = 2500 K) and acetylene-nitrous oxide (T = 3200 K) flames but propane-air (2200 K) and hydrogen-argon flame is used in some special cases. The usual selection of flame for the determination of a given element is shown in Table 10.1. When selecting the flame the detection limits and interferences are taken into account.

Acetylene-nitrous oxide flame is mainly used for the elements with high monoxide dissociation energies which are not atomized properly in other flames. The higher temperature, lower oxygen pressure, and reducing radicals CH and C_2 are responsible for the better atomization.

This flame used on the other has to overcome vaporization interferences from phosphorus, aluminum, and silicon on alkaline earth elements.

The mixture of fuel and oxidant is ignited above the burner. The fast chemical reactions of combustion occur in the primary combustion zone in

Table 10.1   Guide for the Selection of Flame

Propane-air flame	Acetylene-air flame	Acetylene-nitrous oxide flame
Li, Na, K, Rb, Cs	Li, Na, K, Rb, Cs, **Mg, Ca, Sr, Ba,** Cr, Mo, Mn, Fe, Ru, Co, Ni, Cu, Zn, Ga, Rh, Pd, Ag, Cd, In, **Sn,** As, Se, Te, Ir, Pt, Au, Hg, Tl, Pb, Bi	Be, Mg, Ca, Sr, Ba, Sc, Ti, V, B, Al, Si, Ge, Y, Zr, Nb, Mo, Sn, La, Hf, Ta, W, Re, Os

Note: Elements in bold = flame to give the best detection limit.

about 10 μs and the flame gases (N_2, CO_2, CO, H_2O, O_2, H_2, NO, and e^-) reach a high temperature there. There is a hot gas jet formed above the burner with a vertical speed of 10 m/s in acetylene-air flame and this high-temperature, reactive, fast-moving gas is the physical and chemical environment of the sample in the flame atomizer. Temperature, concentration of electrons, oxygen, and radicals (H, OH, CH, CN, C_2, etc.) are the most important factors in a flame. The analytical zone (observation height) is typically at 5- to 15-mm height from the top of the burner and it takes around 1 ms for the sample to reach this point.

The main processes and the different forms of the analyte can be followed in Table 10.2.

The three main consecutive processes in the flame to yield free atoms of the analyte are (1) desolvation, (2) vaporization, and (3) atomization. The time

Table 10.2   Main Processes Taking Place in Flame Atomizer

Place of the process	Process	Form of analyte
Nebulizer	Nebulization	Primary aerosol
Spray chamber	Modification of the aerosol, removal of large droplets, mixing of gases	Secondary aerosol
Flame	Desolvation	Dry aerosol
	Vaporization	molecular gases, metal monoxides, MO
	Atomization	free atoms, M
	MO = M + O	
	Ionization	Metal ions, M^+ + e^-
	Formation of new molecules	MOH, MOH_2

to complete the processes is about 1 ms, so these processes should be extremely fast. The free atom concentration will depend on the number of analyte atoms entering the flame and on the the speed and equilibrium position of the processes as well which depends on the matrix and on the characteristics of the flame. The detailed analytical calibration function of the flame atomizer derived from Equation 3 taking in account the mass balance of the nebulizer spray chamber unit, the change of number of moles, n_{298}/n_T, the thermal expansion of the gases (298/T), and introducing constants to express the efficiency of nebulization η, the efficiency of vaporization ε, and the efficiency of atomization β:

$$A = K_\lambda \, l \, \varepsilon \, \beta \, \frac{q_n \, \eta}{Q_{ox} + Q_f} \, \frac{298 \, n_{298}}{T \, n_T} \, c_s \tag{4}$$

$$K_F = \frac{q_n \, \eta}{Q_{ox} + Q_f} \, \frac{298 \, n_{298}}{T \, n_T} \tag{5}$$

The value of K_f for flame atomizes is around 3×10^{-6} expressing the large dilution of the sample in the atomizer. The parameters in K are mostly instrumental but ε and β depend on the analyte, the matrix, the flame used, and on the fuel-to-oxidant ratio as well. The flames with different fuel-to-oxidant ratio are termed as oxidizing or fuel lean, stoichiometric, and fuel rich or reducing.

Optimization in flame atomic absorption usually means the search for the maximum of K. The most important parameters to be varied are the observation height H and the fuel-to-oxidant flows while Q_{ox} is kept constant. With increasing sample uptake rate q_n the efficiency of nebulization η is reduced and the product of $q_n \, \eta$ has a maximum.

Analyzing real samples one often has to face interferences caused by matrix components. The potential interferences are identified in AA books and cookbooks. Interferences are defined as a difference of the analyte signal measured at the same analyte concentration without matrix and with a given matrix. Interferences can be separated into six general classes: (1) transport interference, INT_{TR}; (2) vaporization interference, INT_{VAP}; (3) dissociation interference, INT_{DISS}; (4) ionization interference, INT_{ION}; (5) distribution interference, INT_{DIST}, and (6) background interference INT_{BG}.

Methods used to minimize or eliminate different interferences are listed in Table 10.3.

Electrothermal Atomization Atomic Absorption Spectroscopy (ETA-AAS)

With electrothermal atomizers, 5 to 20 μl sample solution or a few milligrams of solid sample are deposited in an electrically heated atomizer and gradually heated to produce a transient cloud of atomic vapor to be determined by atomic absorption principle. Most of the electrothermal atomizers are constructed with graphite tube of around 20 to 30 mm length and 5 to 6 mm of inner diameter.

Table 10.3 Methods Used to Eliminate or Minimize Interferences

Method to eliminate or minimize interferences	Type of interference corrected
Choice of flame	INT_{VAP}, INT_{DISS}, INT_{ION}, INT_{DIST}
Optimization of fuel-to-oxidant ratio and observation height	INT_{VAP}, INT_{DISS}, INT_{ION}
Application of releasing agents, $LaCl_3$ + HCl, $SrCl_2$	INT_{VAP}, INT_{DISS}
Application of ionization buffers, CsCl, KCl	INT_{ION}
Matrix matching	INT_{TR}, INT_{VAP}, INT_{DISS}, INT_{DIST}
Background correction	I_{BG}

Electrothermal atomizers with graphite tube are often called graphite furnace (GF-AAS). The graphite tube is inserted between two water-cooled support electrodes usually made of graphite which have the function of current supply and cooling of the tube to ambient temperature after atomization. The supply of current can be (1) from the end of the tube (Figure 10.1a) resulting in a lower temperature at the tube ends and causing memory effects and interferences or (2) from the side of the tube (Figure 10.1b) giving a more isothermal tube without condensation and less interferences.

The graphite furnace is purged with argon gas flow during the operation of the furnace to remove the matrix components and to prevent the tube from oxidation in the high temperature steps. In one group of the furnaces there are separated gas lines for the inner part ($Q_{Ar, in.}$) and for the outer part ($Q_{Ar, out.}$) of the tube as in Figure 10.2a, but in other constructions (Figure 10.2b) there is only a cross flow of argon gas around the tube.

There are different graphite tubes available: electrographite tube, pyrolytically coated electrographite tube, or pyrolytic graphite tube. Pyrolytically coated tubes are the most preferred as the coating makes the graphite tube less porous and less reactive which is very important when refractory carbide-forming elements are analyzed.

The furnace is heated up stepwise functionally in four stages:

1. Drying step (105 to 130°C, 10 to 20 s) — during this stage the solvent is evaporated leaving a solid residue in the tube.
2. Thermal treatment step (350 to 1200°C, 10 to 20 s) — organic matter is ashed, volatile inorganic components are vaporized. The temperature should ideally be set to remove volatile components without any loss of the analyte.

3. Atomization step (1200 to 2800°C, 3 to 7 s) — the pretreated sample is vaporized and atomized to produce the free atom vapor of the analyte. The atomic vapor is produced rapidly as a result of the very fast rise of the temperature and diffuses out of the tube while producing a transient, peak-shape signal. As long as the rate of vaporization and rate of diffusion are constant, the peak height (A_p) and peak area measured by integration (A_i) are proportional to the mass of the analyte in the tube. The peak area is less dependent on the rate of vaporization which is changing with the matrix.
4. Cleaning step (2000 to 2800°C, 2 to 5 s) — used to remove all the residues of the sample, to make the tube ready for the next sample.

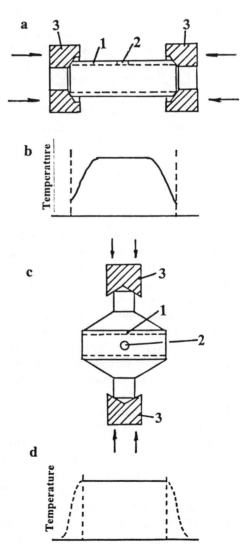

FIGURE 10.1

Heating modes of graphite furnaces: end-heated tube *a* and temperature distribution along the tube *b*, side-heated tube *c*, and temperature distribution along the tube *d*.

a

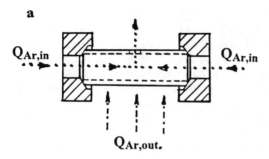

b

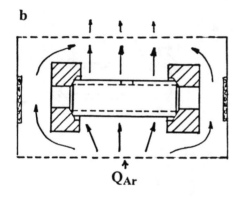

FIGURE 10.2
Purging modes of graphite furnaces: purging with separated inner and outer gas flows *a*, purging with one gas flow *b*.

The programmable power supplies of ETA allow to program around eight to ten program steps and to set the equilibrium temperature, the heating rate (ramp), and the holding time for each step. Further possibilities are (1) to change the gas flow (inner flow) for a given step (air for the ashing step) or to stop or reduce the flow rate during atomization (gas stop, low gas flow rate), (2) to synchronize zeroing, and (3) the measurement of the signal.

The ideal conditions for atomization are (1) high gas temperature, delayed evaporation of the sample and (2) low rate of removal of sample. To ensure these conditions fast heating (2000°C/s), platform or probe atomization and gas stop conditions are recommended. The delay of sample vaporization can be achieved by using a platform, a small piece of flat graphite insertion (Figure 10.3a), curved graphite insertion (Figure 10.3b), a built-in curved platform (Figure 10.3c), or pyrolytic graphite probe (Figure 10.3d) moved in and out synchronized with the program. An analyte sampled on the platform vaporized at the temperature of $T_{pl} = T_{vap}$ will be efficiently atomized in the gas phase since at that time the temperature is higher than the actual temperature of atomization $T_g > T_{at}$, while with a sample on the tube wall the vaporization starts at a temperature $T_w = T_{vap}$ when $T_g < T_{at}$ and the atomization is insufficient. Another good effect of the platform is the localization of the sample to the middle of the tube. The platform temperature is lower than the tube temperature which should be taken into account in the furnace program.

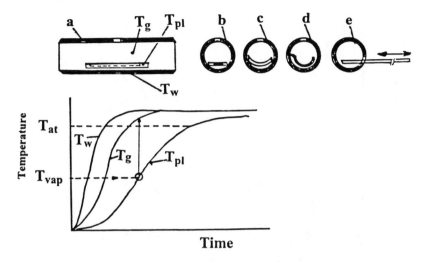

FIGURE 10.3

Principle of platform and probe atomization: positioning of the platform in the graphite tube *a*, view of flat platform *b*, curved platform *c*, built in curved platform *d*, probe atomization *e*. Temperature profile of tube wall T_w, gas phase T_g, and platform T_{pl}.

The conditions of thermal pretreatment can also be critical for highly volatile analytes (As, Cd, Pb, Se, Tl, etc.) as the temperature of the step cannot be set high enough to remove the volatile matrix components without loss of analyte and the residue of the matrix will give a high background absorption. There are routines to overcome this problem: (1) in the sample preparation nitric acid is preferred to hydrochloric acid, chemical modifiers (matrix modifiers) are added to the sample to make the critical matrix component (Cl^-) more volatile (NaCl matrix + NH_4NO_3 modifier) and to make the analyte less volatile. A list of matrix modifiers and the elements affected are given in Table 10.4.

A small volume of liquid sample is introduced into the tube by a microvolume pipette or by an automatic sampling system. The matrix modifier is mixed with the sample prior to sample introduction when manual sampling is applied but this can be done in one step with sophisticated auto samplers.

Table 10.4 Application of Matrix Modifiers

Matrix modifier		Elements affected
$Mg(NO_3)_2$	Mg	Be, V, Cr, Mn, Fe, Co, Zn, Al
$Pd(NO_3)_2 + Mg(NO_3)_2$	Pd/Mg	Cu, Ag, Au, Cd, Hg, Ga, In, Tl, Ge, Sn, Pb, As, Sb, Bi, Se, Te
$Pd(NO_3)_2 + Ca(NO_3)_2$	Pd/Ca	P
$Ca(NO_3)_2$	Ca	B
$NH_4H_2PO_4$		Pb

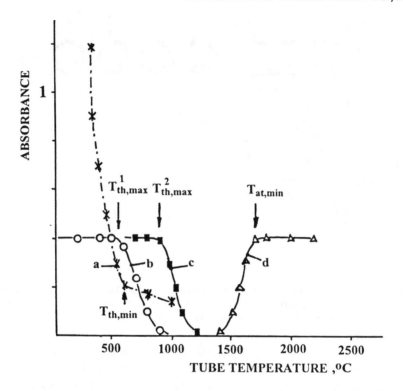

FIGURE 10.4

Thermal pretreatment and atomization curves: background absorbance as a function of thermal pretreatment temperature *a*, analyte absorbance as a function of thermal pretreatment temperature without matrix modifier *b*, analyte absorbance as a function of thermal pretreatment temperature with matrix modifier *c*, atomization curve *d*.

Optimization of the temperature of thermal pretreatment (T_{th}) and atomization (T_{at}) step with sample solution (Figure 10.4): (1) a sample solution giving an absorbance of 0.3 to 0.5 is chosen, (2) the atomization step is set according to manual data, (3) the background absorbance peak is measured and plotted as a function of T_{th} (Figure 10.4a), (4) the analyte absorbance is measured and plotted as a function of T_{th} (Figure 10.4b), (5) the analyte absorbance with matrix modified sample solution is measured and plotted as a function of T_{th} (Figure 10.4c), (6) the analyte absorbance is measured and plotted as a function of T_{at} (Figure 10.4d), and the minimal atomization temperature ($T_{at,min}$) is determined.

From curve *a* the minimal temperature of effective reduction of the background ($T_{th,min}^{Bg}$) is given. Curve *b* and *c* give the maximal pretreatment temperature without ($T_{th,max}^{1}$) and with Pd/Mg matrix modifier ($T_{th,max}^{2}$).

The approximation of the analytical calibration function for ETA-AAS was derived from Equation 3 taking into account the volume of the sample V_s, the volume of the graphite tube V_{TUB}, and the path length l ($V_s = 0.02$ cm^3, $V_{TUB} = 0.8$ cm^3, and l = 3 cm) under optimal conditions:

$$A_p = K_\lambda l \, \epsilon \, \beta \, \frac{V_s}{V_{TUB}} \, c_s \qquad (6)$$

$$K_F = \frac{V_s}{V_{TUB}} \qquad (7)$$

The value of K_F, the dilution factor for electrothermal atomizer, is around 2.5×10^{-2}. As a result of the smaller dilution the sensitivity of the ETA-AAS is typically three orders of magnitude better than the sensitivity of F-AAS; however, in practice the detection limits of ETA-AAS are better by a factor of 10 to 500 than those of F-AAS.

Cold Vapor Atomic Absorption (CV-AAS)

The flame-AAS detection limit of mercury is around 0.2 μg/ml, which is not sufficient for determining environmental levels.

Mercuric ions in solution can be reduced with tin(II)chloride or sodium borohydride to elemental mercury:

$$Sn^{2+} + Hg^{2+} \longrightarrow Sn^{4+} + Hg^\circ$$

$$BH_4^- + 3H_2O + H^+ + 3Hg^{2+} \longrightarrow H_3BO_3 + 3Hg^\circ + 4H_2$$

The atomic mercury vapor is equipartitioned between the aqueous sample and the gas phase. The gas phase is then swept into the long path-length absorption cell constructed with silica windows.

There are two main constructions of mercury accessories used.

(1) In the batch-type instrument (Figure 10.5) first the valves are set to *5b* and *8a* position. The sample (1 to 30 ml) is introduced into the mixed reactor (1), sulfuric acid and tin(II)chloride reagent are added, intense stirring is applied for 40 s and then valve *5* turned to position *a,* the mercury vapor is swept through the drying tube (6) into the cuvette (7), and the transient peak-shaped absorbance signal is recorded. After the signal goes back to zero valve *8* is set to position *b* and the solution is let out into a container and the reactor is rinsed with water. Valve *5* is set then back to position *a.*

(2) In the continuous flow-type instrument (Figure 10.6) the sample and reagent flows are pumped with constant flow rate by a peristaltic pump. First the sample flow (8 ml/min) is mixed with an acid flow (1 ml/min 5 mol/1 HCl), then with the flow of sodium borohydride (1 ml/min, 1% w/v in 0.01% w/v NaOH) and then with inert gas. The mixture passes through the reaction coil (2), then the gas phase of the mixture containing the mercury vapor is separated in the gas-liquid separator (3). The gas is swept through the drying tube (4) into the cuvette and the stationary absorbance level is recorded. The liquid is taken away through a liquid trap to the drain.

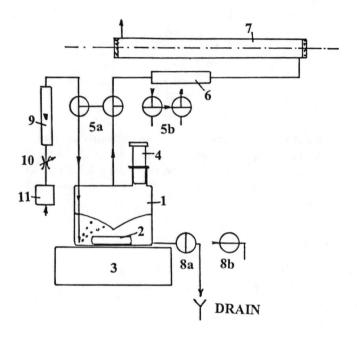

FIGURE 10.5
Drawing of batch-type cold vapor mercury accessory: mixed reactor *1*, stirring bar *2*, magnetic stirrer *3*, insertion port for sample and reagent *4*, double flow three-way valve *5*, drying tube *6*, cuvette with quartz windows *7*, drain valve *8*, flow meter *9*, needle valve *10*, and gas supply *11*.

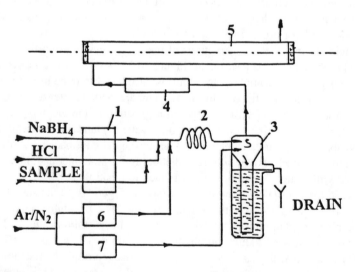

FIGURE 10.6
Drawing of a continuous flow cold vapor mercury accessory: three channel peristaltic pump *1*, reaction coil *2*, gas-liquid separator *3*, dryer tube *4*, cuvette with quartz windows *5*, and gas supplies *6* and *7*.

The continuous flow instrument is favored as it is easier and faster to operate but with the batch-type instrument somewhat better detection limits can be achieved. All the chemicals used should be clean and free from mercury contamination.

10.1.1.   Determination of Iron and Copper in Plant Material by Flame Atomic Absorption Method

Principle of the Determination _____

Before the analysis the plant material (wheat bran, oat bran, etc.) sample is finely ground, homogenized, and dried at 103°C. The sample is dry ashed and the residue is dissolved in acid. The concentration of iron and copper in the solution is determined by flame atomic absorption method and the metal concentration of plant material is calculated.

Apparatus

Atomic absorption spectrometer with acetylene-air flame and computerized data handling
Iron and copper hollow cathode lamp
Volumetric flasks
Pipettes
Adjustable pipette, 1 to 5 ml
Quartz crucible
Electric hot plate with temperature regulation
Electric muffle furnace

Chemicals

Concentrated hydrochloric acid
Concentrated nitric acid
Iron stock solution, 1000 µg/ml Fe
Copper stock solution, 1000 µg/ml Cu

Sample Preparation

Weigh 4 g of sample to the nearest of 0.001 g into a quartz crucible of 50 mm of diameter. Place the crucible onto an electric hot plate and gradually increase the temperature to char until smoking stops. Place the crucible into an electric furnace at 250°C. Increase the temperature up to 550°C in an hour and keep at this temperature for about 3 to 4 hours. Cool the crucible to room temperature and add 1 ml of 1 + 1 nitric acid. Evaporate the residue to dryness on a hot plate then add 10 ml of nitric acid ($HNO_3 + H_2O = 1 + 9$) and redissolve the residue. After cooling to room temperature, filter the solution, collecting the filtrate in a 50-ml volumetric flask, and wash the filter with 2 ml of diluted nitric acid, then with water.

Preparation of Standard Solution

Prepare a mixed stock 300 μg/ml Fe and 200 μg/ml Cu solution by pipetting 15 ml of 1000 μg/ml Fe standard solution and 10 ml of 1000 μg/ml Cu standard solution into a volumetric flask and diluting to 50 ml with water.

Preparation of Iron and Copper Standard Solutions

Code	Concentration of elements Fe (μg/ml)	Cu (μg/ml)	Volume of flasks (ml)	Volume of stock solution 300 μg/ml Fe and 200 μg/ml Cu (ml)	Additive
SD0	0	0	100	0	
SD1	3	2	100	1	20 ml 1 + 9
SD2	6	4	100	2	nitric acid
SD3	9	6	100	3	
SD4	12	8	100	4	
SD5	15	10	100	5	

Procedure

Determination of copper

Wavelength	324.7 nm
Spectral bandwidth	0.5 nm
Maximum lamp current*	15 mA
Working lamp current*	5 mA
Flame	Acetylene air
Flame stoichiometry	Oxidizing
Burner	10 cm acetylene air
Observation height	10 mm
Integration time	2 s
Number of integrations	5

Determination of iron

Wavelength	248.3 nm
Spectral bandwidth	0.2 nm
Maximum lamp current*	20 mA
Working lamp current*	8 mA

---

* Depending on the construction of the lamp.

Flame	Acetylene air, stoichiometric
Flame stoichiometry	Stoichiometric
Burner	10 cm acetylene air
Observation height	10 mm
Integration time	2 s
Number of integrations	5

Put the atomic absorption spectrometer in operation according to the Operational Manual and set the parameters of copper. Aspirate the standard solution SD4 and adjust the burner position to get maximum absorbance.

Load the concentrations of calibration solutions into the computer. Aspirate the standard solutions (SD0-SD5) and carry out the calibration. Calculate the calibration curve using the built-in curve-fitting program then check the calibration data comparing the measured and calculated figures. Print out the calibration data.

Measure the sample solutions five times each and print out the copper concentration of the solution.

Adjust the atomic absorption instrument for the determination of iron by exchanging the hollow-cathode lamp, changing the SBW and wavelength. Carry out the calibration and determine the iron concentration of the sample solution as in the case of copper.

Evaluation

Calculate the copper and iron concentration of dry plant material in mg/kg units from all five solution concentrations measured of copper and iron taking into account the weight of sample and the volume of solution. Calculate the average copper and iron concentrations. Calculate the standard deviations and the relative standard deviations characterizing the precision of the atomic absorption determination of copper and iron under the given conditions. Compare the shape of copper and iron calibration curve.

10.1.2. Determination of Lead in Plant Material by Graphite Furnace Atomic Absorption Method

Principle of the Determination

Before the analysis the plant material (wheat bran, oat bran, etc.) sample is finely ground, homogenized, and dried at 103°C. The sample is dry ashed and the residue is dissolved in acid. The concentration of lead in the solution is determined by graphite furnace atomic absorption method using Pd/Mg matrix modifier and platform atomization.

Apparatus

Atomic absorption spectrometer with graphite furnace,
   background correction, and computerized data handling
Lead hollow cathode lamp
Volumetric flasks
PE container with cover, volume of 1.5 ml
Pipettes
Adjustable pipette of 0.1 to 1 ml
Micropipette, 20 µl
Quartz crucible
Electric hot plate with temperature regulation
Electric muffle furnace

Chemicals

Concentrated nitric acid
Lead stock solution 500 µg/ml Pb in HNO_3
Magnesium nitrate stock solution, 4000 µg/ml Mg
Palladium nitrate stock solution, 4000 µg/ml Pd

Sample Preparation

Weigh 4 g of sample to the nearest of 0.001 g into a quartz crucible of 50 mm diameter. Place the crucible onto an electric hot plate and gradually increase the temperature to char until smoking stops. Place the crucible in an electric furnace at 250°C. Increase the temperature up to 550°C in an hour time and keep at this temperature for about 3 to 4 hours. Cool the crucible to room temperature and add 2 ml of 1 + 1 nitric acid. Evaporate the residue to dryness on a hot plate then add 10 ml of nitric acid ($HNO_3 + H_2O = 1 + 9$) and redissolve the residue. After cooling to room temperature, filter the solution, collecting the filtrate in a 50-ml volumetric flask, and wash the filter with 2 ml of diluted nitric acid, then with water.

Preparation of Lead Standard Solutions

Pipette 0.75 ml of lead stock solution (500 µg/ml Pb) with adjustable pipette into a 100-ml volumetric flask, add 10 ml of 1 + 9 nitric acid. Using this solution prepare the solutions below:

CODE	Concentration of Pb (ng/ml)	Volume of flask (ml)	Volume of stock solution (3.75 µg/ml Pb, ml)	Additive 1 + 9 HNO_3 (ml)
Pb0	0	25	0.0	5
Pb1	30	25	0.2	5
Pb2	60	25	0.4	5
Pb3	90	25	0.6	5
Pb4	120	25	0.8	5
Pb5	150	25	1.0	5

Preparation of Matrix Modified Standard Solutions and Samples

Mix 1 ml palladium nitrate solution (4000 µg/ml Pd) with 1 ml magnesium nitrate solution (4000 µg/ml Mg) to prepare Pd/Mg matrix modifier additive.

Pipette 1-1 ml of standard solution and sample into small PE containers. Add 0.2 ml of mixed Pd/Mg matrix modifier solution then mix the solutions thoroughly. (As long as the dilution of samples and standards with the matrix modifier is the same, the original concentrations can be used in the calibration procedure.)

Procedure

Wavelength	283.0 nm
Spectral bandwidth	0.5 nm
Maximum lamp current*	15 mA
Working lamp current*	5 mA
Background correction	ON
Sample aliquot	20 µl
Graphite tube	Electro-graphite with platform
Purging gas	Ar

Furnace Program

Step	1	2	3	4	5	6
Temperature (C°)	150	800	20	2300	2500	20
Ramp time (s)	1	1	1	0	1	1
Hold time	20	20	10	4	2	15
Read				X		
Baseline			X			
Flow (ml/min)	100	100	100		100	100
Mini flow						
Stop flow				X		

Put the atomic absorption spectrometer in operation according to the Operational Manual and set the parameters for lead. Set the graphite furnace in position and adjust the optical position. Load the the graphite furnace program. Run test measurements with the standard solution Pb5 and with the sample solution. Evaluate the signal shape and position, the level and structure of background, and set the start and end of peak integration. Carry out the calibration by measuring the peak height and peak area absorbance with the

---

* Depending on the construction of the lamp.

calibration solutions. Calculate the calibration curve using the built-in curve-fitting program then check the measured and calculated figures and the curves on the screen. Repeat the measurement of the deviating calibration points and recalculate the calibration curve if it is necessary. Print out the calibration data.

Inject 20 µl of the sample solution five times and print out the lead concentration of the solution obtained in the individual measurements.

Evaluation

Calculate the lead concentration of dry plant material from the solution concentrations measured taking into account the weight of sample and the volume of solution. Calculate the average concentration in mg/kg units. Calculate the standard deviation and the relative standard deviation characterizing the precision of the atomic absorption determination of lead under the given conditions.

10.1.3. Determination of Mercury in River Sediment Using Cold Vapor Atomic Absorption Technique

Principle of the Determination

Toxic heavy elements such as mercury are usually enriched in the sediment of rivers and ponds. Since mercury and its organic derivatives are poisonous even when present in small amounts, a method of general utility became necessary.

Before the analysis the river sediment is homogenized and dried at ambient temperature. The sample is acid digested in PTFE pressure bomb then the concentration of mercury is determined with batch or continuous flow cold vapor atomic absorption technique.

Apparatus

> Atomic absorption spectrometer with background correlation and computerized data handling
> Cold vapor mercury accessory
> Mercury hollow cathode lamp
> PTFE bombs
> Volumetric flasks
> Pipettes
> Adjustable pipette 1 to 5 ml

Chemicals

> Concentrated nitric acid
> Hydrogen peroxide, 30 m/v%
> Mercury stock solution, 1000 µg/ml
> Magnesium perchlorate

Sample Preparation

Weigh 1, 1-g of sample portions on analytical balance into three PTFE bombs (use quartz insert if it is available). Add 3 ml concentrated nitric acid and 1 ml hydrogen peroxide. Close the bombs and digest the samples at 140°C for 3 hours. Cool down the bombs to room temperature. Open the bombs and filter the solution, collecting the filtrate in 50-ml volumetric flasks, and wash the filter with 2 ml diluted nitric acid, then with water. The solutions are coded *S1, S2, S3*.

Blank solution is prepared by dilution of 4 ml concentrated nitric acid to 100 ml and coded *BL*.

Procedure

Wavelength	253.7 nm
Spectral bandwidth	0.5 nm
Maximum lamp current*	10 mA
Working lamp current*	3 mA
Background correction	ON

Procedure of Mercury Determination with Batch-Type Cold Vapor Accessory

Solutions

20% w/v $SnCl_2$ solution, 100 ml
Dissolve 23.8 g $SnCl_2\ 2H_2O$ in 80 ml 1+1 HCl, warm gently if necessary. Add 1+1 HCl to get 100 ml solution.
1+4 H_2SO_4, 100 ml
20 ml of concentrated H_2SO_4 is poured carefully into 80 ml water.
Mixed reagent, 100 ml
Mix 50 ml of 20% w/v $SnCl_2$ solution with 50 ml of 1+4 H_2SO_4.

Put the atomic absorption spectrometer in operation according to the instructions for use. Install the batch mercury cold vapor accessory and insert a freshly prepared drying tube.

Pour 10 ml of sample or standard solution into the mixed reactor of the mercury accessory then add 2 ml of mixed reagent and close the reactor. Mix the solution for 40 s then start the signal reading and the gas flow at the same time. Read the peak height absorbance. Wait until the signal returns to zero then let the solution out into the waste container. Rinse the reactor and set the valves ready for the next measurement.

---

* Depending on the construction of the lamp.

Preparation of the Standard Solutions

Code	Concentration of mercury (mg/l)	Volume of flask (ml)	Volume of stock solution (1 ng/ml Hg, ml)	Additive
HgB0	0.00	100	0	
HgB1	0.01	100	1	
HgB2	0.02	100	2	4 ml
HgB3	0.03	100	3	1+1 HNO$_3$
HgB4	0.04	100	4	
HgB5	0.05	100	5	

Load the concentrations of the standard solutions into the computer, then measure the standard solutions one by one, adding 10-10 ml of solutions (HGB0 to HGB5) and 2 ml of mixed reagent into the reactor. Apply the curve-fitting program to calculate the calibration curve. Check the errors of the calibration comparing the measured and calculated figures. Repeat the measurement of the deviating standard points and run the curve-fitting routine again. Store and print out the data of calibration.

Determine the mercury concentration of the sample solutions BL, S1, S2, and S3 measuring three times each solution in triplicate using 10 ml of aliquot and 2 ml of mixed reagent following the procedure above.

Procedure of Mercury Determination with Continuous Flow Cold Vapor Accessory

Chemicals

> 1+1 HCl
>
> NaOH
>
> NaBH$_4$

Solutions

1% w/v NaBH$_4$ in 0.1% w/v NaOH, 200 ml
> Dissolve 0.2 g NaOH in 150 ml water then add 2 g NaBH$_4$ and dissolve. Add water to get 200 ml solution. The solution should be then filtered into plastic storage vessel.

1+1 HCl, 200 ml
> 100 ml concentrated. HCl is poured into 100 ml water.

Preparation of Standard Solutions

Code	Concentration of mercury (mg/l)	Volume of flask (ml)	Volume of stock solution (2 ng/ml Hg, ml)	Additive
HgC0	0.00	100	0	
HgC1	0.02	100	1	
HgC2	0.04	100	2	2 ml
HgC3	0.06	100	3	1+1 HNO$_3$
HgC4	0.08	100	4	
HgC5	0.10	100	5	

Put the atomic absorption spectrometer in operation according to the instructions for use. Install the continuous flow cold vapor accessory and insert a freshly prepared drying tube. Determine the stabilizing time necessary to reach the equilibrium signal. Set the stabilizing time and integration time 5 to 10 s.

Load the concentrations of the standard solutions into the computer, then measure the standard solutions (HGC0 to HGC5). Apply the curve-fitting program to calculate the calibration curve. Check the errors of calibration comparing the measured and calculated figures. Repeat the measurement of the deviating standard points and run the curve-fitting routine again. Store and print out the data of calibration.

Determine the mercury concentration of the sample solutions BL, S1, S2, and S3 measuring three times each solution.

Evaluation

Calculate the mercury concentration of the sample from each measured concentration in mg/kg units. Calculate the average concentration. Evaluate the standard deviation and relative standard deviation.

10.2. Flame Emission Spectroscopy

Principle of the Technique

Flame emission spectroscopy (FES) is an emission spectroscopic technique in which thermal energy liberated in the process of combustion is applied to atomize and to excite the sample delivered into the flame in the form of anaerosol. FES corresponds to electron transitions from low to higher energy state of the free atoms and stable molecules (MO, MOH) in ground state.

Excitation is mainly a thermal type: the atoms or molecules get into excited state by collision with gas molecules of the flame. Part of the exited atoms, M* loses the excitation energy (E_j) by collision; however, another part returns to the ground state by emitting the excitation energy by emission of a photon of given frequency v:

$$M^* \longrightarrow M + h\nu$$

The excitation energy and wavelength of the emitted light is characteristic of the nature of the atoms and molecules. The emitted intensity I_{em} of a spontaneous emission line in a simplified situation can be given as

$$I_{em} = A_{ij}\, h\nu_{ij}\, N \frac{g_j}{g_o}\, e^{-\frac{E_j}{kT}} \tag{8}$$

where A_{ij} is the transition probability, h is the Planck constant, v_{ij} is the frequency of the line, N is the total number of atoms in the observed zone of flame which is directly related to the concentration of the solution, g_o and g_j are the statistical weight of the ground and excited states, respectively, k is a constant, and T is the temperature of the flame. The intensity of atomic emission is highly dependent on the temperature; however, under the conditions of flame emission measurements all the parameters in Equation 1 are constant except N. Thus, Equation 1 becomes

$$I_{em} = K_e\, N = K_e\, K_F\, c_s \tag{9}$$

where K_e is the constant of the excitation and emission process, K_F is the constant of the sample introduction and the flame atomization process, c_s is the concentration of analyte in the sample solution (see also Section 10.1, Equations 4 and 5).

When low concentrations of analyte atoms are used (i.e., when self-absorption is negligible), the emission intensity is proportional to the sample concentration. At higher sample concentrations, if self-absorption is of appreciable degree, the calibration curve declines.

Flame photometers are frequently used for alkali metal determinations in clinical and agricultural analyses. In such instruments low-temperature flame (e.g., propane air) is used to excite the most prominent lines and the lines are isolated by interference filters. Simultaneous multielement instrumentation (Na, K, Li) is commonly used.

Most of the flame emission work is performed on atomic absorption instruments used in emission mode as these instruments are more available in the laboratories. With the use of high-temperature nitrous oxide-acetylene flame, it is possible to extend flame emission method to a higher number of elements.

Flame background and molecular emission cause particular problems of flame emission spectroscopy. The commonly used on-peak background correction by the subtraction of blank reading from the reading obtained when the analyte solution aspirated into the flame can be a source of errors as the sample solution may alter the background. In background-limited cases the off-peak background correction is the best solution but this method is usually not available in the instruments.

10.2.1. Determination of Sodium in Plant Material by Flame Emission Method

Principle of the Determination

Before the analysis the plant material (wheat bran, oat bran, etc.) sample is finely ground, homogenized, and dried at 103°C. The sample is dry ashed and then the residue is dissolved in acid. The concentration of sodium in the solution is determined by flame atomic emission method and the sodium concentration of plant material is calculated.

Apparatus

> Flame emission spectrometer or atomic absorption
>   spectrometer in emission mode with acetylene-air flame
>   and computerized data handling
> Volumetric flasks
> Adjustable pipette 1 to 5 ml
> Quartz crucible
> Electric hot plate with temperature regulation
> Electric muffle furnace

Chemicals

> Concentrated hydrochloric acid
> Sodium stock solution, 1000 µg/ml
> Potassium stock solution, 50 g/l

Sample Preparation

Weigh 4 g of sample on analytical balance into a quartz crucible of 50-mm diameter. Place the crucible onto an electric hot plate and gradually increase the temperature to char until smoking stops. Place the crucible in an electric furnace at 250°C. Increase the temperature up to 550°C in an hour time and keep at this temperature for about 3 to 4 hours. Cool the crucible to room temperature and add 2 ml of 1 + 1 nitric acid. Evaporate the residue to dryness on a hot plate, then add 10 ml of $1 + 9$ nitric acid ($HNO_3 + H_2O = 1 + 9$) and redissolve the residue. After cooling to room temperature filter the solution, collecting the filtrate in a 50-ml volumetric flask, and wash the filter with 2 ml of diluted nitric acid, then with water.

Preparation of Sodium Standard Solutions

Code	Concentration of Na (µg/ml)	Volume of flasks (ml)	Volume of stock solution (200 µg/ml Na, ml)	Additive
NaSD0	0	100	0	
NaSD1	2	100	1	20 ml 1 + 9
NaSD2	4	100	2	nitric acid and
NaSD3	6	100	3	4 ml 50 g/l K
NaSD4	8	100	4	
NaSD5	10	100	5	

Procedure

Wavelength	589.0 nm
Spectral bandwidth	1 nm
Flame	Acetylene air
Flame stoichiometry	Stoichiometric
Burner	Maker burner or
	10 cm acetylene-air burner set
	right angle to the optical beam
Observation height	10 nm
Integration time	2 s
Number of integrations	5

Put the flame spectrometer or atomic absorption spectrometer (emission mode) in operation according to the Instructions for use and set the parameters for sodium. First aspirate the standard solution NaSD5 of 10 µg/ml Na and adjust the intensity reading to maximum by changing the gain, then aspirate the solution NaSD0 of 0 µg/ml Na and set the baseline to zero. Check the maximum reading again and repeat the procedure if necessary. Load the concentrations of calibration solutions into the computer. Carry out the calibration by reading the intensities while aspirating the standard solutions. Calculate the calibration curve using the built-in curve-fitting program. Check the calibration by comparing the measured and calculated figures. Print out the calibration data.

Measure the sample solution with five repeated readings. Dilute the sample solution if the reading is outside of the range and repeat the measurement.

Evaluation

Calculate the sodium concentration of dry plant material in mg/kg units from the solution concentration readings taking into account the weight of sample and the volume of solution. Calculate the average concentration, the standard deviation, and the relative standard deviation as well.

10.3. Inductively Coupled Optical Emission Spectroscopy

Principle of the Technique _____

The method of inductively coupled plasma optical emission spectroscopy (ICP-OES) is based on the measurement of the optical emission of free atoms or ions excited in an inductively coupled plasma source. The sample is introduced into the plasma in the form of wet or dry aerosol by argon gas flow where all the components of the sample are vaporized, atomized, and ionized. The free atoms and ions are excited and the line spectra of the given atoms and ions are emitted from the plasma. After the separation of one wavelength with a monochromator (one line) or a number of wavelengths at a time with a polychromator from the spectra, the line intensities are measured by an optical detector. The quantitative determination of the elements is done after calibration with standard solutions.

The analytical calibration function is principally linear

$$I_A = k_c c \tag{10}$$

where I_A is the line intensity of analyte, c is the concentration of the analyte in the solution analyzed, and k_c is the constant of the calibration function. However, at high analyte concentrations the slope of the calibration function generally decreases as a result of self absorption and instrumental nonlinearities.

The inductively coupled plasma (ICP) discharge is suspended by a high-frequency (27.17 MHz, 40.68 MHz) magnetic field by the transfer of energy into a flowing ionized gas, usually argon. A schematic drawing of a typical ICP is given in Figure 10.7. The so-called "torch" is usually an assembly of three concentric tubes made of quartz. The top section of the torch is positioned inside the water-cooled, two- or three-turn induction coil of a radio frequency generator. The outer argon flow (10 to 15 l/min) sustains the high-temperature plasma, gives the position of the plasma relative to the outer walls and the induction coil, and prevents the walls from melting. The plasma under these conditions has an annular shape. The sample aerosol carried with the inner argon flow (0.5 to 1.5 l/min) enters the central channel of the plasma. The intermediate argon flow (0 to 1.5 l/min) is optional and has the function of

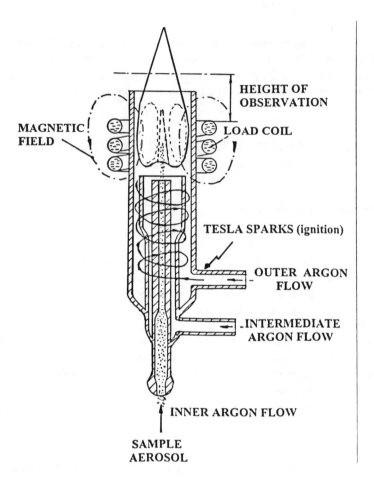

FIGURE 10.7
Schematic drawing of ICP.

lifting up the plasma slightly and diluting the inner gas flow in the analysis of organics.

The performance characteristics of the inductively coupled argon plasma emission method can be given as follows:

1. Temperature in the observation zone 7000 to 8000 K
2. High electron concentration ($n_e = 10^{14}$–10^{16} cm^{-3})
3. High level of ionization for most of the elements
4. Low background emission
5. 70 elements can be determined including phosphorus and sulfur
6. Elements difficult to measure with AAS spectroscopy can be measured, e.g., B, C, Ce, La, Nb, P, Pr, S, Ti, V, W, and Zr
7. Linear calibration function over four to six orders of magnitude of concentration

8. Small vaporization and ionization interference
9. Spectral interferences (line overlap) may be one of the limiting factors of the method
10. Simultaneous multielement capability
11. Detection limits are in the range of 0.1 to 100 ng/ml
12. Precision 0.2 to 2% RSD

The most common technique of sample introduction is to nebulize solutions pneumatically or ultrasonically using nebulizers of different construction (concentric nebulizer, cross-flow nebulizer, V-groove nebulizer, grid nebulizer, ultrasonic nebulizer, etc.). However, there are a number of special techniques for the introduction of liquids, slurries or solids (laser ablation, spark aerosol generation, electrothermal vaporization), and gases (hydride generation) as well.

The complete ICP-OES instrument consist of a (1) radiofrequency generator, (2) a coupling unit, (3) a plasma unit with the induction coil, torch, gas supplies and flow controllers, and sample introduction apparatus (usually nebulizer and spray chamber), (4) imaging optics, (5) a monochromator or multichannel spectrometer with off-peak background correction, and (6) a readout and control unit and data processing and display unit.

The spectra of the sample constituents are superimposed on the background emission of the plasma. The background of the plasma fed with water can be considered as the minimum background consisting of a continuum, the atomic spectra of argon, hydrogen, oxygen, and the molecular spectra of OH, CN, C_2, N_2^+, NO, and CO. The matrix components can also alter the structure and level of the minimum background which can be treated as spectral interference.

The basic types of backgrounds are shown in the form of wavelength scans in four pairs of figures. The first scans are for the analyte without matrix; the second scans are for the analyte with a given matrix. Figure 10.8 *a1*, *a2*, *b1*, *c1*, and *d1* represents simple background, *b2* a sloping background, *c2* a complex background caused by partial spectral overlap, and *d2* a complex background caused by direct spectral overlap.

The intensity of analyte line (I_A) is determined using background correction by Equation 11:

$$I_A = I_{AB} - I_B \qquad (11)$$

where I_{AB} is the total intensity, I_B is the background intensity, all intensities defined for the wavelength of the analysis line.

On-peak background correction can be applied when the structure and level of background is constant and a suitable reference blank solution can be prepared. The background intensity I_B is determined by the measurement of reference blank solution on the analysis line; then this background is applied in Equation 2.

The simple background level change (Figure 10.9a) can be corrected by one off-peak measurement on either side of the line

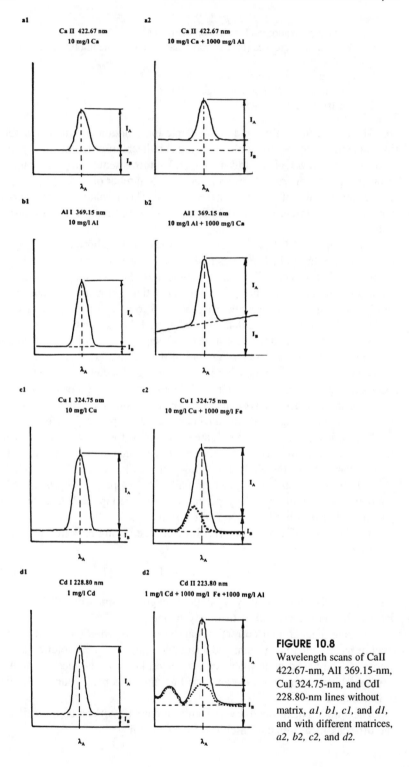

FIGURE 10.8
Wavelength scans of CaII 422.67-nm, AlII 369.15-nm, CuI 324.75-nm, and CdI 228.80-nm lines without matrix, *a1, b1, c1,* and *d1,* and with different matrices, *a2, b2, c2,* and *d2.*

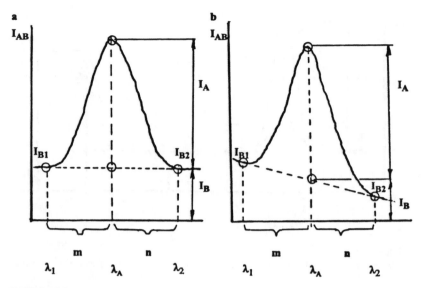

FIGURE 10.9

Correction of single background a and sloping background b using off-peak background correction principle.

$$I_A = I_{AB} - I_{B1} = I_{AB} - I_{B2} \qquad (12)$$

To correct a sloping background (Figure 10.9b), two measurements are needed at a distance of a and b from the peak

$$I_A = I_{AB} - \frac{b\, I_{B1} + a\, I_{B2}}{m + n} \qquad (13)$$

In a general case (Figure 10.10) there is a parallel or sloping background shift and a direct spectral overlap with the line of an identified element. In this case the off-peak background correction can be combined with interelement correction based on the use of λ_1 and λ_2 lines of the interfering element. The actual intensity I_1 of the λ_1 line determined from the measured intensity I_2 of λ_2 line applying the intensity ratio is determined in a calibration step. The line intensity ratio of the interfering element is

$$k_i = \frac{I_1}{I_2} \qquad (14)$$

Then the intensity of the analyte line with off-line background correction and interelement correction is

$$I_A = I_{AB} - I_{B1} - I_1 = I_{AB} - I_{B1} - k_i\, I_2 \qquad (15)$$

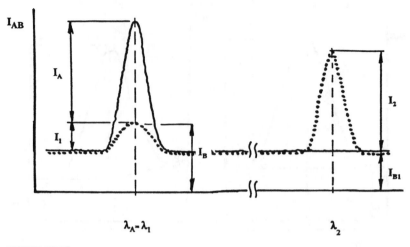

FIGURE 10.10

Treatment of complex background using off-peak background correction and interelement correction.

$$I_A = I_{AB} - \frac{b\,I_{B1} + a\,I_{B2}}{m + n} - k_i\,I_2 \qquad (16)$$

The background correction is one of the most critical steps of the ICP-OES measurement which needs a careful study of the lines.

To determine the analyte concentration from the measured analyte line intensity, the analytical evaluation function is used:

$$c = k_e\,I_A \qquad (17)$$

where k_e is the constant of the evaluation function.

The ICP-OES method is generally used as a simultaneous or sequential multielement technique. Depending on the concentration range used a linear or nonlinear calibration routine is applied.

Calibration on the linear section of the evaluation function is carried out by two-point calibration using a blank solution and one or a number of multielement calibration solutions containing all the elements measured in known concentration. From the concentration and intensity data pairs two constants, the slope and the intercept of the evaluation function, are calculated and stored for all the elements. (With efficient off-line background correction the value of the intercept is near zero.) Slight changes of the instrumental parameters with time can alter the value of the two constants and a concentration error can occur. The error is eliminated by recalibration.

To cover the whole analytically useful concentration range the nonlinear part of the evaluation function should be used as well. In such a case blank and five to seven standard solutions of different concentrations for each element are measured under stabilized conditions, then the evaluation functions of each element are determined by polynomial curve fitting. The polynomial evaluation function is

$$c = a + bI_A + cI_A^2 + dI_A^3 \qquad (18)$$

where a, b, c, and d are constants determined in the first full range calibration step.

As the curvature of the evaluation functions represented by constants c and d changes only little as a result of drift, it is common to use a simplified, two-point recalibration routine measuring a blank and one standard solution for each element on the linear section of the calibration function. The actual values of a and b are then calculated with linear approximation.

10.3.1. Determination of Mineral Components of Plant Material Using Simultaneous Multielement ICP-OES Method

Principle of the Determination

Before the analysis, the air-dried plant material sample (leaves, grass, wheat bran, oat bran, etc.) is finely ground, homogenized, and dried at 103°C. The sample is acid digested in PTFE pressure bombs, then the concentrations of the elements are determined by simultaneous multielement ICP spectrometer.

Apparatus

ICP spectrometer
PTFE bombs
Volumetric flasks
Pipettes
Adjustable pipette 1 to 5 ml

Chemicals

Concentrated nitric acid
Hydrogen peroxide 30 m/v %
Element stock solutions, 1000 µg/cm^3
10 µg/cm^3 manganese solution (Mn1)

Sample Preparation

Weigh 0.5 g of sample portions on analytical balance into two PTFE bombs. Add 3 cm^3 concentrated nitric acid and 1 cm^3 hydrogen peroxide. Close the bombs and put them into the heater block set to the temperature of 140°C. Digest the sample at this temperature for 3 h then cool down the bombs to room temperature. Open the bombs and filter the solution, collecting the filtrate in one 50-ml volumetric flask, and wash the filter with 2 ml of diluted nitric acid then with water. The solution is marked with code S1.

Blank solution is prepared by the dilution of 8 ml concentrated nitric acid to 100 ml and marked by code BL.

Apparatus

Spectrometer	Pashen-Runge, F = 1 m, vacuum
Plasma	Argon-argon
Torch	Fassel type
Frequency	27.17 MHz
Incident power	1.2 kW
Reflected power	<5 W
Argon flow rates	
Outer	12 1/min
Intermediate	1 1/min
Inner	0.8 1/min
Observation height	13 mm
Nebulizer	V-grove, GMK nebulizer
Sample flow rate	2.7 cm^3/min
Integration time	5 s
Delay time	20 s
Washout time	25 s

Procedure

Switch on the ICP spectrometer then ignite the plasma according to the instruction manual. Select the optimal instrumental parameters. Wait 30 min to stabilize the plasma and optics then check the optical alignment of the ICP spectrometer and the sample introduction: (1) check and set the true wavelength of the polychromator; nebulize the 10 µg/ml manganese solution (Mn1) and run a wavelength scan covering the manganese line profile; check the line position; calculate the wavelength drift and reset the wavelength scale of the polychromator if necessary. (2) Run a time-resolved measurement with Mn1 solution and check the rise and drop of the intensity, the noise and fluctuation of the signal, and determine the delay time and washout time.

Load the previously developed and tested multielement measurement program capable of measuring 16 elements documented in Table 10.5. The program involves off-peak background correction and interelement correction to correct for the interference from iron as well (* for elements corrected). Detection limits of the chosen lines defined in solution concentration are given in column 5 of Table 10.5, while detection limits calculated for the original sample form with a multiplication factor of 50 are in column 6. The multiplication factor is given by the sample preparation (f = 50 ml/1 g = 50 ml/g).

Table 10.5   Line Selection, Standard Solutions, and Detection Limits for Solution and for Sample

Code	Element/line (nm)		Concentration (µg/ml)	Acid	Limit of detection solution (µg/ml)	Limit of detection sample (µg/g)
SD1	Ca (I)	422.67	100	5% v/v HNO$_3$	0.006	0.3
	K (I)	766.49	100		0.2	10
	Mg (II)	279.55	5		0.0002	0.01
	Na (I)	589.59	100		0.16	8
	Sr (II)	407.77	5		0.0004	0.02
SD2	*Cd (I)	228.8	5	5% v/v HNO$_3$	0.001	0.05
	*Co (I)	238.89	5		0.005	0.25
	*Cr (II)	205.55	5		0.006	0.3
	*Cu (II)	224.70	5		0.01	0.5
	Fe (II)	259.94	5		0.008	0.4
	Mn (II)	257.61	5		0.002	0.1
	Ni (II)	231.5	5		0.008	0.4
	*Pb (II)	220.35	5		0.04	2
	Zn (I)	213.94	5		0.006	0.3
SD3	P (I)	178.20	100	5% v/v HNO$_3$	0.04	2
	S (I)	180.73	100		0.03	1.5
BL			0	5% v/v HNO$_3$		

Run the recalibration routine of the program starting with the blank solution BL, followed by the multielement standard solutions SD1, SD2, and SD3, and reset the analytical evaluation functions. Choose sample measurement from the program and first measure the blank and standard solutions as samples to check the proper operation of the instrument and program. The blank readings should be near zero or in the order of detection limit while the readings for standard solution should be the nominal value with an accuracy of 2 to 3%.

Measure the sample solutions doing five replicate measurements loading the sample code and the dilution factor, f (f = V/m, V is volume of solution, 50 ml, m = 1g is the mass of sample) into the computer before the measurement to calculate the element concentrations in the original sample. Use diluted sample solution for the elements whose concentration exceeds the calibrated range of the program.

Evaluation

Tabulate the element concentrations of the sample calculated from the replicate measurements (c1, c2, c3, c4, c5, µg/g) according to Table 10.6, then calculate the mean concentrations, $\overline{c}$, the sample standard deviations, s, and coefficient of variation (CV = $\overline{c}/s \times 100\%$) for each element. Round the mean concentrations ($\overline{c}_R$, µg/g) using the significant-figure convention taking into account the values of sample standard deviation. Use the data processing program of the ICP spectrometer or a spreadsheet program for these calculations.

Table 10.6

Element	c1 (µg/g)	c2 (µg/g)	c3 (µg/g)	c4 (µg/g)	c5 (µg/g)	$\overline{c}$ (µg/g)	$\overline{c}_R$ (µg/g)	s (µg/g)	CV (%)

Load rounded concentrations, $\overline{c}_R$, variation coefficients, CV, from Table 10.1 and the detection limits valid for the original sample, c_L, from Table 10.5 into Table 10.7. Calculate

Table 10.7

Element	$\overline{c}_R$ (µg/g)	c_L (µg/g)	$\overline{c}_R/c_L$	CV (%)

the $\overline{c}_R/c_L$ ratios for each element. Plot the variation coefficients as a function of log($\overline{c}_R/c_L$) in the format given below using data of all elements.

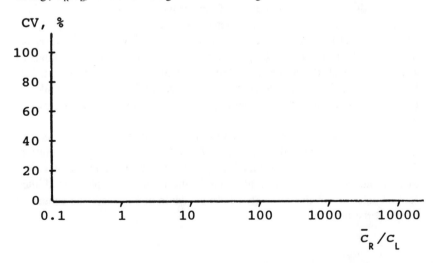

10.4.   References

Boumans, P. W. J. M., *Inductively Coupled Plasma Emission Spectroscopy,* Vol. I and II, John Wiley & Sons, New York, 1987.

Ebdon, L., *An Introduction to Atomic Absorption Spectroscopy — A Self Teaching Approach,* Heidon, Philadelphia, 1982.

Ingle, J. D. and Crouch, S. R., *Spectrochemical Analysis,* Prentice-Hall, New York, 1988.

Ottaway, J. and Ure, A. *Practical Spectrometric Analysis,* Pergamon Press, Elmsford, NY, 1983.

Weltz, B., *Atomic Absorption Spectroscopy,* Verlag Chemie, New York, 1976.

X-Ray Diffraction
Techniques

11.1. Identification of the Components of a Sample with the Aid of the Diffractogram

Principle of the Technique

X-ray diffraction methods are suitable for the examination of the crystalline materials. A part of the radiation incident upon the crystal scatters on the crystal lattice. The scattered radiation can be well observed only in directions in which the beams reflected from the crystal planes under each other are amplified by interference. According to Bragg, in real crystals with an interplanar spacing (lattice constant) *d,* which contain a great number of planes, the reflection is detectable only at given incident angles, namely, some of the X-rays are transmitted inside the crystal and are subsequently reflected from succeeding parallel planes (Figure 11.1). The reflected beams, because the difference in the distances traveled by the rays /s'-s/ is about the same magnitude as the wavelength, interfere.

According to Bragg, the reflected beams give detectable maxima only when their differences in the distances traveled are equal to the integral multiple of wavelength (λ). In quantitative interpretation of this phenomenon, Bragg instead of the incident angle used its complementary angle, i.e., the reflection angle (θ), which is known as the Bragg angle. A radiation with wavelength λ is reflected from a set of planes with *d*-spacing only at conditions, when

$$\eta\lambda = 2d_{hkl} \sin \theta$$

where n is an integer, hkl is the Miller index of the plane.

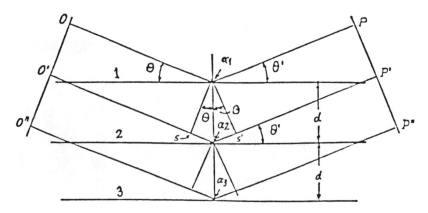

FIGURE 11.1

Diffraction of X-rays on the crystal lattice according to Bragg (d — interplanar spacing, θ - Bragg angle).

This relation known as the Bragg equation, in spite of the fact that it is derived by a high degree of simplification of the diffraction phenomena, is the most frequently used equation in X-ray diffraction examinations.

Measuring System

X-ray Generator

The X-ray generator consists of an X-ray tube and equipment for the regulation and stabilization of the intensity of the beam (high-voltage transformer, measuring instruments, stabilizer).

X-ray Tube

For the generation of X-ray radiation usually a tube with hot filament cathode is used. The voltage of the tube depends on the material of anticathode (or anode) and varies from 30.000 to 80.000 V.

Filter

The emitted beam consists of a continuum (white radiation) and some characteristic wavelengths. The latter depends on the target element of the anode. For the diffraction measurements monochromatic X-rays are needed, which can be achieved by

1. Using an absorption edge filter
2. Using a single crystal monochromator placed before or after the sample
3. Energy discrimination.

When using an absorption edge filter, a sheet of metal foil is placed in the path of the beam whose absorption edge is between the K_α and K_β radiation. In this way the intensity of K_β lines and those of the continuous radiation are decreased. (The atomic number of the element in the filter generally is one or two less than that of the target material, e.g., $Cu_{(anode)} - Ni_{(filter)}$, $Co_{(anode)} - Fe_{(filter)}$, etc.)

The intensity of the disturbing radiation can be decreased by a monochromator. A set of planes of a single crystal reflects in accordance with the angle of incidence and angle of reflection only the radiation with the wavelength (n = 1) and its overtones (n > 1) determined by the Bragg equation. The beam obtained after the monochromator crystal is not only monochromatic but its divergence is rather small. However, the loss of intensity in the monochromator is very high; it reflects only 1 to 2% of the incident radiation. To increase the intensity of the monochromatic beam, various curved-crystal monochromators are developed (Figure 11.2).

Sample Holders
The sample holder can be made of glass, plastic, or light metals.

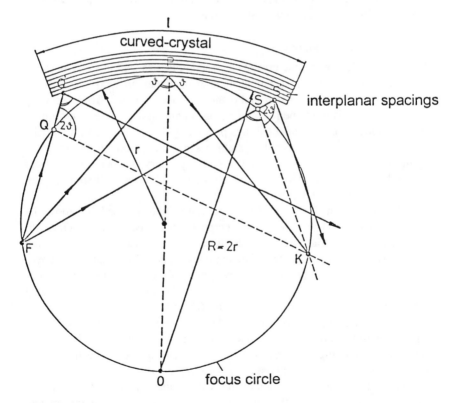

FIGURE 11.2
Curved-crystal monochromator with Johansson focusing optic.

Detectors

The detectors of the reflected radiation of a polycrystalline material can be arranged into two groups:

1. The reflected beams are recorded at the same time in the whole angle range on a sensible photographic film. For this purpose different chamber types (Debye-Scherrer, Guinier) are used.
2. The different reflections are detected successively, only at a given dispersion angle, by a suitable detector (Geiger-Müller, proportional, semiconductor, scintillation counting tube). This kind of detection is usually used in powder diffractometers.

Powder Diffraction Methods

From the analytical point of view, among the X-ray diffraction methods, the powder method has the most significant practical importance. The method was developed almost at the same time by Debye-Scherrer in Germany and Hull in the U.S. By the powder method all of the crystalline materials can be examined, large crystals as well as powdered materials.

The Principle of the Method

In a fine powder sample the crystals are randomly arranged in every possible orientation and there is a sufficient number of planes oriented at the Bragg angle, θ; thus, for these crystals and planes diffraction occurs. The possibility of the reflection is the same for all the planes lying in the path of the primary beam. It means that the reflected rays pass along the superficial of a 2θ half-angle cone. If a fine-powdered crystalline sample is placed in the axis of a cylindrical Debye-Scherrer chamber, and the primary monochromatic beam enters the chamber radially, the reflected beam reaches the photographic film as shown in Figure 11.3.

For the examination a fine-powdered sample must be used that allows its even dispersion over the whole plane perpendicular network space. The level of disorder of the orientation of crystals can be increased by rotation of the sample, so in this way sharper diffraction lines appear. The height of a Debye-Scherrer chamber is a few centimeters; thus, only the reflections with very small or very large angles give a whole circle on the film; from other lines only the central section is recorded by the film (Figure 11.4).

The evaluation of the record is based on the determination of the reflection angles, 2θ, of the individual lines that allows the calculation of the interplanar distances, d, causing the reflection by the Bragg equation. The interplanar distances, $d_1...d_n$ are characteristic for the sample.

If $r = 28.7$ mm is the radius of the X-ray camera, then $I = \theta$, i.e., in the evaluation of the record, the arc length values I measured in millimeters directly give θ, the reflection angle according to Bragg, thereby the evaluation of the records is rendered very convenient. Consequently such radian-diameter

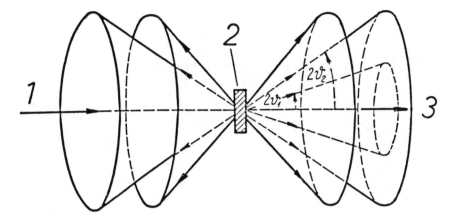

FIGURE 11.3
Diffraction of X-rays on a powder sample. 1: incident beam; 2: powder sample; 3: reflected beam.

cameras are very widely used. Larger cameras are made preferably so as to have a radius that is a multiple of the above value (Figure 11.5).

In the Debye-Scherrer method, for the detection of the X-ray a photographic film is used. In diffractometers, detectors are applied instead of the film. The goniometer of a diffractometer is working on the basis of Bragg-Brentano's focusing principle (Figure 11.6). The X-ray is emerging from the

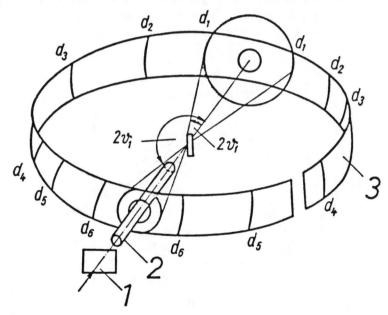

FIGURE 11.4
Schematic representation of the geometry used in the Debye-Scherrer technique. 1: filter; 2: entrance slit system; 3: film.

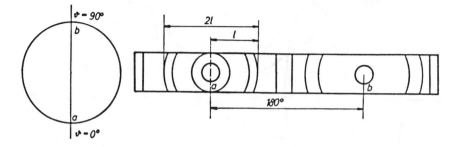

FIGURE 11.5
Film mounting according to Straumanis.

splitting diaphragm, B_1, or from a focus line that is perpendicular to the paper sheet in the point F of the figure. The divergence of the radiation is regulated by a diaphragm B_2. A plane polycrystal sample (P) is placed by its plane surface rotatable in the center (O) of a measuring circle. The detector is rotating on the circumference of the measuring circle by a double angular velocity related to the sample. The incident direction and the reflection direction always constitute an identical angle with the plane perpendicular. In this way the beam reflects from every position of the sample with an angle 2θ. The reflected beams are focused in the point B_3, where the detector is placed.

    The evaluation of the record is based on the determination of the reflection angles, 2θ, of the individual lines. By the Bragg equation, from the determined Bragg angles one can calculate the interplanar distances, d, causing the reflection.

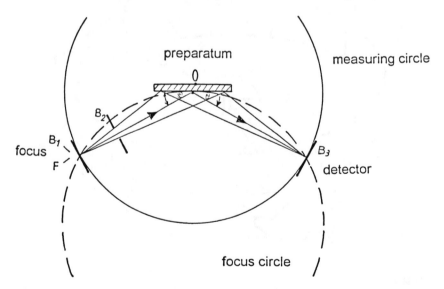

FIGURE 11.6
Goniometer geometry according to the Bragg-Brentano principle.

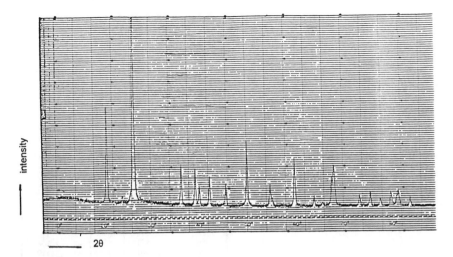

FIGURE 11.7
The diffractogram of α-quartz.

Identification of the Components of a Sample with the Aid of the Diffractogram

The result of the evaluation of a diffractogram of an unknown sample is a set of interplanar spacings, d_{hkl}, with the corresponding intensities (Figure 11.7). On the basis of these data one can determine the composition of the sample, that is, one has to find a substance or substances whose set of diffraction lines are identical with those of the sample.

For the identification of the components of a sample one can use the JCPDS data base (on cards or on CD-ROM). The JCPDS card system is supplemented by a four-book register (Figure 11.8). In the first and second book, the organic and the inorganic substances are listed in alphabetical order by their English name. After the name the d values of the three most intensive lines with their intensities are given followed by the catalog number of the JCPDS card with the appropriate complete data system. In the third book, the substances are classified in the "Hanawalt" system on the basis of the d value of their most intensive line into 45 groups. Inside a group the lines of a substance are listed with decreasing intensity. In the fourth book the substances are listed on the basis of FINK system. Here 1 of the 4 most intensive lines is on the first place, and adequate to this the substance is listed in 1 of the 45 groups. The other d values are listed by decreasing intensity. Each substance is listed four times in the book, choosing each line from the four most intensive ones as definitive.

Task

Record the diffractogram of two or three samples by powder diffraction technique. The evaluation of the recorded diffractogram is

d	3.34	4.26	1.82	4.26	SiO2
I/I1	100	35	17	35	SILICON DI OXIDE ALPHA QUARTZ

Rad. Cu   λ 1.5405   Filter
Dia.   Cut off   Coll.
I/I1   d corr. abs.?
Ref. SWANSON AND FUYAT, NBS CIRCULAR 539, /ol. III (1953)

Sys. HEXAGONAL   S.G. D_3^4 - $C3_12$
a_0 4.913 b_0 c_0 5.405 A   C1.10
α β γ Z 3
Ref. IBID.

εα   nωβ 1.544ε γ 1.553   Sign +
2V D_x 2.647 mp   Color
Ref. IBID.

MINERAL FROM LAKE TOXAWAY, N.C. SPECT. ANAL.:
<0.01% AL; <0.001% CA,CU,FE,MG.
X-RAY PATTERN AT 25°C.

REPLACES 1-0649, 2-0458,2-0459, 2-0471, 3-0419, 3-0427, 3-0444

d Å	I/I1	hkl	d Å	I/I1	hkl
4.26	35	100	1.228	2	220
3.343	100	101	1.1997	5	213
2.458	12	110	1.1973	2	221
2.282	12	102	1.1838	4	114
2.237	6	111	1.1802	4	310
2.128	9	200	1.1530	2	311
1.980	6	201	1.1408	<1	204
1.817	17	112	1.1144	<1	303
1.801	<1	003	1.0816	4	312
1.672	7	202	1.0636	1	400
1.659	3	103	1.0477	2	105
1.608	<1	210	1.0437	2	401
1.541	15	211	1.0346	2	214
1.453	3	113	1.0149	2	223
1.418	<1	300	0.9896	2	402,115
1.382	7	212	.9872	2	313
1.375	11	203	.9781	<1	304
1.372	9	301	.9762	1	320
1.288	3	104	.9607	2	321
1.256	4	302	.9285	<1	410

FIGURE 11.8
A JCPDS card.

$$d = \frac{\lambda}{2} \cdot \frac{1}{\sin \theta}$$

For the copper anode tube $\lambda_{CuK\alpha} = 1.5405$ Å; for the chromium anode tube $\lambda_{CrK\alpha} = 1.7900$ Å. The identification of the sample is on the basis of the determined d values.

11.2. References

Glocker, L. R., *Materialprüfung mit Röntgenstrahl,* Springer-Verlag, Berlin, 1971.

Henry, N. M. F., Lipson, H., and Wooster, W. A., *The Interpretation of X-Ray Diffraction Photographs,* Macmillan, New York, 1960.

Klug, H. P. and Alexander, L. E., *X-Ray Diffraction Procedures,* John Wiley & Sons, New York, 1959.

Preiser, H. S., Rooksby, H. P., and Wilson, A. S. C., *X-Ray Diffraction by Polycrystalline Materials,* The Institute of Physics, London, 1960.

Part III

Thermal Methods
of Analysis

Chapter 12

Derivatography

Principle of the Technique ─────────────────────────

Derivatography is a compound, dynamic thermoanalytical technique according to which by the aid of a special instrument — the derivatograph — the following parameters, brought about by heating, are measured simultaneously in the sample: change in weight (thermogravimetry, TG curve), rate of the change in weight (derivative thermogravimetry, DTG curve), change in enthalpy (differential thermal analysis, DTA curve), and the temperature of the sample (T curve). The instrument automatically records the above curves as a function of temperature or time. The records thus obtained enable qualitative and quantitative analysis to be carried out.

12.1. Investigation of Calcium Oxalate Monohydrate by Means of the Derivatograph

Principle of the Determination ─────────────────────

Most compounds undergo various physical (modification, etc.), and chemical changes (thermal decomposition, etc.) on heating. These changes are always connected with an enthalpy change, and some of them with weight change, gas evolution, and change in volume as well. Thermoanalytical methods are based upon the measurement of these changes.

The derivatograph is a complex thermoanalytical equipment. With this device the weight change (TG curve), the rate of weight change (DTG curve), and the temperature change (T curve) of a single sample as well as the temperature difference between sample and a reference material can be measured simultaneously; the latter is proportional to the enthalpy changes of the sample (DTA curve). With the help of special adapters joined to the derivatograph, the process of gas evolution as well as the thermodilatation of the sample can also be measured simultaneously.

The equipment records the variables mentioned on a photosensitive chart in the form of curves, as functions of time.

Apparatus

> Derivatograph, MOM type
> Analytical balance

Chemicals and Utensils

> Calcium oxalate monohydrate precipitate
> Aluminum oxide (ignited at 1200°C)
> Photographic paper
> Developing bath
> Fixing bath
> Drying equipment

Procedure

Put the derivatograph into operation according to the directions for use.

Adjust a heating rate of 10°C/min and a final temperature of 1000°C.

Weigh 700 to 730 mg of calcium oxalate monohydrate into the platinum crucible of the derivatograph, and the same amount of aluminum oxide into another identical crucible. Place the crucibles into the apparatus.

Record the thermal curves and after developing the photographic chart, evaluate them as follows.

Prepare the Temperature Calibration Lines and Inscriptions

Fix the diagram onto a drawing board. Draw parallel lines through the intersection points of the T curve and the temperature calibration lines at 20°C intervals. Draw the lines belonging to 100, 200, 300,...°C more markedly. Supply the diagram with the appropriate numerical values belonging to the time, weight, and temperature calibration lines and the curves with the characteristic temperature values.

Qualitative Analysis

By investigating composite systems of unknown composition (e.g., bauxite), the aim is in general to detect and identify the individual components. This is based on the comparison and verification of the characteristic transformation temperatures (peak temperatures).

The purpose of investigating pure substances is in general to establish what kinds of transformations and at what temperatures take place in the compound, what is the course of the transformation processes, and how and to what extent the experimental conditions (e.g., presence of oxygen) affect the transformation itself.

In the case of calcium oxalate investigation, the thermal decomposition of the sample took place in the known way according to the following equations:

$$CaC_2O_4 \cdot H_2O \xrightarrow{\text{max } 250°C} CaC_2O_4 + H_2O \uparrow$$

$$CaC_2O_4 \xrightarrow{\text{max } 470°C} CaCO_3 + CO \uparrow$$

$$CaCO_3 \xrightarrow{\text{max } 880°C} CaO + CO_2 \uparrow$$

If we compare the decomposition temperatures of these decomposition processes, as known from the literature, with the peak temperatures of the curves, we find that the values are in good agreement with one another when the heating rate is low.

From the DTA curve it can also be seen that the first and third decomposition processes were of endothermal character, while in the course of the second decomposition process an endothermal and an exothermic process took place simultaneously. Accordingly, these latter two processes overlapped each other, so that the specially shaped section of the DTA curve around 470°C can be regarded as the resultant of two processes of opposite direction. The decomposition of anhydrous calcium oxalate is obviously an endothermal process, so it can be assumed (or even proved) that the exothermic effect was caused by combustion of carbon monoxide liberated in the course of the reaction.

Quantitative Analysis

After interpreting the course of the curves and identifying the individual processes, we can begin the quantitative evaluation of the diagram.

First determine on the DTG curve those temperatures at which the previous process ended, but the next one has not yet started. Project these points onto the TG curve. On the basis of the weight changes observed in these temperature intervals and with the knowledge of the original weight of the sample, carry out stoichiometric calculations. State the amount of the individual components of a composite system (e.g., hydrargillite, boehmite, etc. in bauxite).

In the present case the composition of the sample is known. However, the aim of the examination may also be to find out the extent to which the amount of the decomposition products, their composition, and the course of the reaction differ from the theoretical.

Now, in the knowledge of the amount of the sample and decomposition products, examine in weight percents to what extent the measured values differ from the theoretical ones.

1.   Amount of CaC_2O_4 (theoretical: 87.6%)

$$x_1 = \frac{w_{(II)}}{w_{(I)}} \times 100\% \, CaC_2O_4$$

2. Amount of $CaCO_3$ (theoretical: 68.5%)

$$x_2 = \frac{w_{(III)}}{w_{(I)}} \times 100\% \, CaCO_3$$

3. Amount of CaO (theoretical: 38.4%)

$$x_3 = \frac{w_{(IV)}}{w_{(I)}} \times 100\% \, CaO$$

4. Amount of H_2O

$$x_4 = \frac{w_{(I)} - w_{(II)}}{w_{(I)}} \times 100\% \, H_2O$$

5. Amount of CO (theoretical: 19.2%)

$$x_5 = \frac{w_{(II)} - w_{(III)}}{w_{(I)}} \times 100\% \, CO_2$$

6. Amount of CO_2 (theoretical: 30.1%)

$$x_6 = \frac{w_{(III)} - w_{(IV)}}{w_{(I)}} \times 100\% \, CO_2$$

where $w_{(I)}$ is the weight of the original sample, $w_{(II)}$ is the weight of the sample after the first decomposition process, $w_{(III)}$ is the weight of the sample after the second decomposition process, and $w_{(IV)}$ is the weight of the sample after the third decomposition process.

The peak areas of the DTA curve are proportional to the amount of heat evolved or absorbed during the transformations. Accordingly, we may determine the approximate value (with an accuracy of 5 to 10%) of the transformation heat if we calibrate the instrument under the very same circumstances.

In the present case the problem is to determine the reaction heat of the dehydration process:

$$CaC_2O_4 \cdot H_2O = CaC_2O_4 + H_2O \uparrow + Q_1$$

Instead of calibrating, start from the fact that the reaction heat of the process:

$$CaCO_3 = CaO + CO_2 \uparrow + Q_2$$

($Q_2 = 43,750$ cal/mol) is known.

Put a transparent paper upon the diagram and trace the first and third endothermal peaks of the DTA curve. After cutting out the two peak areas, weigh the paper slips on an analytical balance. The weights of the paper slips will be proportional to their surface area. The dehydration heat can be calculated with the help of the following equation:

$$Q_1 = \frac{g_2}{g_1} \cdot \frac{V_2}{V_1} \cdot Q_2 \ \text{cal/mol}$$

where Q_1 is the dehydration heat, Q_2 is the dissociation heat, g_1 is the area under the first peak, i.e., the weight of the paper slip, g_2 is the area under the third peak, i.e., the weight of the second paper slip, V_1 is the thermal voltage of the thermocouple corresponding to 50°C (0.421 mV) in the neighborhood of the first peak (250°C), and V_2 is the thermal voltage of the thermocouple corresponding to 50°C (0.555 mV) in the neighborhood of the third peak (880°C).

It is to be noted that the quotient $\frac{V_2}{V_1}$ must be incorporated into the equation, because the thermal voltage of the thermocouple does not increase linearly with the temperature; accordingly, different peak areas belong to the same amount of heat at 250 and at 880°C, respectively. The results are approaching the theoretical ones, when the investigated amount of sample is small.

12.2.   References

Paulik, J. and Paulik, F., Simultaneous thermoanalytical examinations by means of the derivatograph, in *Wilson and Wilson's Comprehensive Analytical Chemistry,* Vol. 12A, Svehla, G., Ed., Elsevier, Amsterdam, 1981.

Chapter **13**

Thermo-Gas Titrimetry

Principle of the Technique

Very accurate quantitative determinations can be carried out on the basis of the TG and DTG curves obtained with the derivatograph, provided that only one decomposition product is produced from the sample upon the action of heat. If two or more gaseous decomposition products are liberated, the TG or DTG curves show only the sum of the two decomposition processes. If it is necessary to determine the individual decomposition products selectively and continuously, the thermo-gas titrimeter apparatus, which can be connected to the derivatograph as an adapter, can advantageously be used. In thermo-gas titrimetric measurements, the gaseous decomposition products produced upon the action of heat are absorbed in an absorbent liquid of adequate composition and are quantitatively determined by some sort of titration technique (e.g., acid-base, redox, or precipitation). An adequately chosen pair of electrodes is immersed into the absorbent liquid; the electrodes monitor by a change in their potential if the composition (e.g., pH) of the absorbent solution is changed upon the action of the decomposition product. Delivery of the titrant is automatically started in the titration apparatus upon the action of the change in the potential of the electrode. The rate of the addition of the titrant is controlled by the difference between the measured voltage signal and a reference voltage. By means of a voltage recorder it is possible to continuously record the amount of the titrant added and, what is equivalent to the latter, the amount of material absorbed, as a function of temperature or of time. The curve thus obtained is the thermo-gas titrimetric (TGT) curve. The thermo-gas titrimeter also allows recording the derivated TGT curve, the DTGT curve.

13.1. Recording the TG, DTG, DTA, TGT, DTGT, and T Curves of Cupric Tetrammine Sulfate Monohydrate with a Derivatograph Equipped with a Thermo-Gas Titrimeter, in an Atmosphere of Air

Principle of the Determination _____

It can be seen in Figure 13.1, showing the derivatogram of cupric tetrammine sulfate monohydrate, that the first DTG peak is a double one, whereas only one peak can be observed on the DTGT curve. It follows from these facts that the change in weight in this range is a composite process, that is, crystal water and ammonia escape simultaneously. The DTGT curve shows only one of the two decomposition products: ammonia.

The determination of the ammonia content is carried out by acid-base titration with a glass and saturated calomel electrode pair and with 0.1 N hydrochloric acid as titrant. The absorbent solution is distilled water containing 3% ethyl alcohol whose pH has been adjusted to 5.0 with hydrochloric acid.

Apparatus

Derivatograph equipped with a thermo-gas titrimeter
pH meter
Glass electrode
Saturated calomel electrode

Chemicals

Cupric tetrammine sulfate monohydrate distilled water
containing 3% ethyl alcohol (absorbent solution)
0.1 N hydrochloric acid titrant

Procedure

The apparatus is put into operation and 200 cm^3 absorbent solution, whose pH was adjusted to 5.0, is poured into the absorber prior to starting the titration. (The same solution will be used to rinse the gas analysis tube.)

If the pH of the absorbent solution deviates by 0.1 pH from the adjusted value upon the action of the decomposition product absorbed, delivery of the titrant is automatically started and is carried on as long as the pH is 5.0 again. The apparatus records the required amount of titrant as a function of time.

Evaluation

Draw the TG, DTG, DTA, and TG curves plotted against temperature. Calculate the amount of the decomposition product titrated on the basis of the TGT

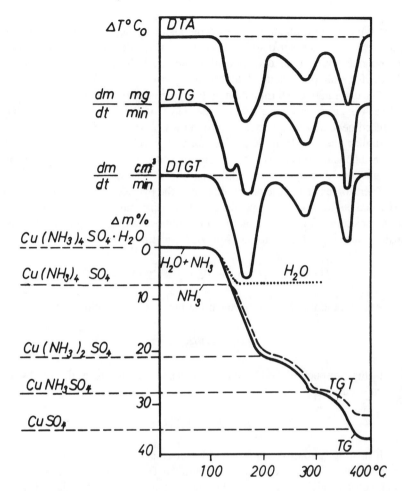

FIGURE 13.1
Derivatogram of copper tetrammine sulfate.

curve in milligram units and determine the composition of the various phases formed in the course of the decomposition. (1 cm^3 of the 0.1 N HCl solution is equivalent to 1.703 mg NH_3.)

13.2. Determination of the Pyrite Content of Bauxite by the Thermo-Gas Titrimetric Technique

Principle of the Determination

The main mineral components of bauxites (e.g., hydrargyllite, gothite, diaspor, caolinite, etc.) are decomposed with the liberation of water upon heating. The determination of the quantity of the decomposition products on the basis of the TG curve presents no difficulties. However, bauxites often contain additional components which, upon decomposition, yield CO_2, SO_2, or SO_3. Such minerals are, for example, calcite, dolomite, and pyrite.

The quantitative determination of the pyrite content of bauxite on the basis of the TG curve is difficult, since the wave of the small amount of SO_2 produced generally totally merges into the water wave. The case of dolomite is a similar one, in which the product formed is CO_2.

In such and similar cases, the thermo-gas titrimetric (TGT) technique can advantageously be applied to supplement the TG technique.

Thermo-gas titrimetric determination of the pyrite content of bauxites can be carried out in two ways, depending on the experimental circumstances of the decomposition of the bauxite sample. In an atmosphere of oxygen, the decomposition occurs according to the following equation:

$$2FeS_2 + 6 1/2 O_2 = Fe_2O_3 + 2SO_2 + 2SO_3$$

whereas in an atmosphere of nitrogen the process is the following:

$$FeS_2 = FeS + S$$

In the presence of Fe_2O_3 and an atmosphere of nitrogen, the following pattern occurs:

$$FeS_2 + 5Fe_2O_3 = 11FeO + 2SO_2$$

If heating is controlled in such a manner that SO_2 and SO_3 are produced simultaneously, the task can be solved with the TGT technique, with an acidi-alkalimetric measurement. If the pH of the absorbent solution is adjusted to 5.0 and hydrogen peroxide is added to it as an oxidizing agent, both SO_2 and SO_3 can be titrated, in the form oxidized to sulfuric acid, with an alkali as titrant.

If the decomposition is conducted in such a manner that only SO_2 is produced, its amount can be measured oxidimetrically with the TGT technique and the pyrite content of the bauxite can be determined. In the latter case, the SO_2 formed is absorbed in a strongly acidic solution and titrated with a solution of $KMnO_4$. A Pt electrode is used as the indicator electrode and a saturated calomel electrode as reference.

Apparatus

> Derivatograph equipped with a thermo-gas titrimetric adapter
> pH meter
> Pt indicator electrode
> Saturated calomel reference electrode

Chemicals

> Pyrite-containing bauxite sample
> 0.1 M potassium permanganate titrant solution
> Absorbent solution: 200 mg $MnSO_4$ and 5 cm^3 concentrated
> H_2SO_4 are dissolved in 300 cm^3 distilled water

Procedure

Depending on its pyrite content, a 100- to 500-mg portion of the bauxite sample is weighed into the large-size sample holder of the derivatograph on an analytical balance.

Prior to starting the titration, 300 cm^3 of the absorbent solution is poured into the absorber. Introduction of the carrier gas and suction through the absorber are started. (The rate of the former should be 20 to 25 l/h, that of the latter 8 to 10 1/h.)

Hereupon 0.1 cm^3 0.1 N KMnO$_4$ solution is added manually to the absorbent solution and the signal source of the reference voltage is brought to such a position that the potential difference between the electrodes should be equal to the reference voltage. This state is reached if the pilot light on the titrator goes out and remains permanently dark.

The burette is now switched over from manual to automatic delivery. In the following, the burette will automatically deliver the titrant at a rate corresponding to the rate of SO_2 addition to the absorbent solution.

After these preliminary operations the apparatus is put into operation as described in the instructions manual, the recorder and the heating program controller are switched on, and the TG, TGT, and T curves are recorded.

Evaluation

Calculate the amounts of SO_2 (in milligram units) equivalent to the amounts of titrant required up to ten points chosen on the strongly sloping portions of the curve. Draw the results obtained into the original thermogram, in accordance with its scale of change in weight (1 cm^3 0.1 N KMnO$_4$ titrant corresponds to 3.2 mg SO_2).

Chapter 14

Thermometry

Principle of the Technique _____

Every chemical transformation is accompanied by changes in enthalpy and under adiabatic conditions by changes in temperature. This well-known phenomenon is the basis of thermometry.

Since the amount of the substance formed in the course of a reaction is always defined by the reaction partner being present in a lower concentration, the reaction heat — and, under appropriate conditions, also the temperature change — will be proportional to this.

The amount of heat formed in the system is

$$Q = n_m \Delta H \tag{1}$$

where n_m is the mole number of the reacting component and ΔH is the molar enthalpy change.

According to the fundamental correlation known from calorimetry:

$$Q = k \Delta T \tag{2}$$

where k is the heat capacity of the system.

From Equations 1 and 2:

$$\Delta T = n_m \frac{\Delta H}{k} \tag{3}$$

ΔH is constant at constant pressure and in a small temperature interval. If the heat capacity of the system (volume of the test solution, its specific heat, etc.) is always the same, the value of k is also constant; so, under adiabatic conditions

$$\Delta T = k' n_m \tag{4}$$

i.e., the change in temperature is proportional to the mole number of the component investigated. The slope k' is the greater the reaction heat and the smaller the heat capacity of the system.

14.1. Thermometric Determination of Active Chlorine in Chlorinating Materials

Apparatus with Accessories

Directhermom apparatus
Pipettes, 2 cm^3, 5 cm^3, 10 cm^3, 20 cm^3

Chemicals

Commercial chloride of lime, active chlorine content 25 to 30%
Commercial sodium hypochlorite "Hypo," chlorine content 90 to 150 g/l
Pharmaceutical benzene sulfochloramide sodium "Neomagnol," chlorine content 27 to 29%
Saturated chlorine water, active chlorine content 0.15 to 0.20 g/l
"Thermo-neutral" potassium iodide (90 cm^3 20% potassium iodide + 10 cm^3 cc hydrochloric acid) 0.1 N sodium thiosulfate solution

Procedure

The solutions used for the measurements and the instrument must be of the same temperature.

Preparation of the Calibration Curve

Three parallel measurements are carried out in each case with different amounts of chlorine water. In the first experiment 2 cm^3, then 5 cm^3, finally 10 cm^3, and 20 cm^3, respectively, of the chlorine water are added to the distilled water in the plastic beaker of the Directhermom. The distilled water is of room temperature and its amount is calculated so that by adding to it the amount of chlorine water in question it should be completed exactly to 200 cm^3. A magnetic stirrer is put into the beaker which is brought into the instrument.

Fill the 2 cm^3 immersion pipette with the reagent, clamp it into the measuring head, and dip into the solution. Start stirring; wait 3 min until thermal equilibrium is established. Adjust the galvanometer at sensitivity 1 × to 0, then add the reagent to the solution and read the deflection of the galvanometer (20 to 30 s).

Because of the volatility of chlorine water, determine the amount of the chlorine present in the reaction space during the reaction by the measurement of the equivalent amount of elementary iodine liberated from the reagent. For this purpose, the plastic beaker is taken out of the instrument and put upon a magnetic stirrer motor, then the iodine is titrated with 0.1 N sodium thiosulfate solution. The deflection is plotted as the function of the chlorine in milligrams determined in the 200-cm^3 test solution.

Determination of the Active Chlorine Content of "Hypo"

Make a stock solution of the "Hypo" by a 20-fold dilution. Then draw 5-, 10-, 15-, and 20-cm^3 aliquots and complete to 200 cm^3 with distilled water. Pour the solution into the plastic beaker of the instrument. Add 2 drops of 10% hydrochloric acid for neutralization. Carry out the measurement as described earlier. (The time of maximum deviation is 20 s.) From the calibration line the content of chlorine can be read and controlled by the titration of the iodine. Compare with the standard!)

Determination of the Active Chlorine Content in Chloride of Lime

Put the plastic beaker containing 200 cm^3 of distilled water into the instrument. Add 0.1 to 0.2 g of the chloride of lime (weighed with 0.1-mg accuracy). After stirring the solution for 3 min, neutralize by adding 2 drops of 10% hydrochloric acid. Finally, carry out the measurement in the way described earlier. (The maximum deflection can be read only after 100 s. This is probably the reason for the 1.5% deviation, originating from the calibration curve.)

Determination of the Active Chlorine Content in Neomagnol

Break the approximately 1-g tablet into five to ten parts and use 0.1- to 0.2-g particles for one measurement. After weighing the sample throw it into the beaker containing 200 cm^3 of distilled water. Fill the 2-cm^3 immersion pipette with thermo-neutral potassium iodide solution. While the solution is stirred for 3 min as usual, the tablet will be dissolved and also thermal equilibrium will be established. Carry out the measurement as described earlier.

A 17% deviation can be observed with respect to the calibration curve taken earlier, because the chlorogene is accompanied in the Neomagnol tablet by 8% binding material causing side reactions. Therefore, a new calibration line must be prepared on the basis of titrating the iodine liberated in the course of the process with 0.1 N sodium thiosulfate solution.

14.2. Checking the Thermal Effects of Reagents

Principle of the Determination

The reagents used are of high concentration and small volume in order to have the thermal equilibrium between sample solution and reagent established speedily. When the concentrated reagent solutions are mixed with the more dilute sample solution, the mixture warms up or cools down on account of hydration. To avoid this, indifferent additives are admixed to the reagents until thermal compensation is attained.

If, for example, concentrated hydrofluoric acid is applied as a reagent, the solution warms up; let the loss in heat be Q_1. However, if urea is dissolved in the hydrogen fluoride solution, a loose complex compound is formed which dissociates upon dilution: it takes up a heat quantity of Q_2. By applying the urea in an appropriate proportion, $Q_1 + Q_2 = 0$. It was found empirically that for this purpose 150 g urea is to be dissolved in 1000 cm^3 40% hydrogen fluoride solution.

Apparatus

Directhermom apparatus
Plastic beakers, 300 cm^3
Immersion pipettes
Magnetic stirrer rod
Beckman thermometer
Burettes, 25 cm^3
Graduated cylinder, 20 cm^3
Pipette, 10 cm^3

Chemicals

Hydrochloric acid solution, specific gravity 1.12
Saturated mercuric chloride solution
Hydrogen fluoride solution containing urea (150 g urea dissolved in 1000 cm^3 40% hydrogen fluoride)

Procedure

10 cm^3 of the HgCl$_2$ solution is added to 60 cm^3 hydrochloric acid solution and the volume is made up to 200 cm^3. The temperature of the mixture is adjusted to room temperature. The mixture is poured into plastic beakers and placed into a Dewar flask. The 5-cm^3 immersion pipette (for hydrogen fluoride) is filled with urea-containing hydrogen fluoride solution and immersed, together with the calorifer and the thermistor, into the sample solution. Stirring is started. The system is given a waiting time of about 3 min to let the temperatures of the sample and of the reagent solutions equalize. Switch "THERM" is switched on and the galvanometer is adjusted to scale division 200. The hydrogen fluoride is poured into the solution to be tested. If the temperature of the solution is increased upon the addition of the reagent, some more urea is dissolved in the hydrogen fluoride, whereas, if the temperature is decreased, a few cubic centimeters of hydrogen fluoride is added and the deflection of the galvanometer is repeatedly checked.

This procedure is continued until no galvanometer deflection is observed.

14.3. Determination of the Fe₂O and SiO₂ Content of Concrete from a Single Weighing In

Principle of the Determination _____

Ferric ions in the concrete sample dissolved in hydrochloric acid are reduced to ferrous ions with sodium polysulfide and the excess sulfide is removed from the solution with mercuric chloride. Hereupon the ferrous ions are oxidized to ferric ions with hydrogen peroxide and the increase in temperature — which is proportional to the iron oxide content — is measured. Hydrogen fluoride is added to the solution which forms a hexafluoro complex with silicic acid. The change in temperature brought about by the heat of reaction is proportional to the silicic acid content.

Apparatus

> Directhermom apparatus
> Plastic beakers, 300 cm^3
> Immersion pipettes
> Magnetic stirrer rod
> Beckman thermometer
> Burettes, 25 cm^3
> Graduated cylinder, 20 cm^3
> Pipette, 10 cm^3

Chemicals

> Hydrochloric acid solution, specific gravity 1.12
> Saturated sodium polysulfide solution
> Saturated mercuric chloride solution
> Hydrogen peroxide solution (100 cm^3 water + 100 cm^3 33% H$_2$O$_2$ + 28 cm^3 concentrated H$_2$SO$_4$)
> Hydrogen fluoride solution containing urea (150 g urea dissolved in 1000 cm^3 40% H$_2$F$_2$)

Procedure

2.00 g of the finely pulverized concrete sample is weighed into a dry 600-cm^3 glass beaker. 100 cm^3 water is added and the mixture is vigorously stirred with a magnetic stirrer in order to prevent sedimentation of the concrete. With continued stirring, a 60-cm^3 portion of HCl is added and stirred for further 3 min. Na$_2$S$_x$ solution is dropped to the mixture as long as it becomes colorless (about 0.5 to 1 cm^3). 10 cm^3 of the HgCl$_2$ solution is added and the volume of the solution is made up in a graduated cylinder to 200 cm^3. The temperature of the solution is adjusted to room temperature; it is poured into a plastic beaker and placed into the Dewar flask. The pipettes are filled: the one used for hydrogen fluoride with 5 cm³ hydrogen fluoride solution, the other with 1 to 2 cm³ H_2O_2. The pipettes, the thermistor, and the calorifer are immersed into the solution. The magnetic stirrer is started and the solution is given about 3 min of waiting for the temperatures to equalize.

The shunt calibrated for the determination of the SiO_2 content is switched on, the galvanometer is zeroed in setting "THERM.", and the H_2F_2 is brought into the solution.

The reaction comes to an end in about 20 to 30 s, as observed on the maximum deflection of the galvanometer. The value read from the scale directly gives the SiO_2 content in percent.

Hereupon the shunt serving for the determination of Fe_2O_3 is switched into circuit, the galvanometer is repeatedly zeroed, and the H_2O_2 is brought into the sample solution. The maximum of the galvanometer deflection, indicating completeness of the reaction, is read.

Evaluation

Calculate the Fe_2O_3 and SiO_2 content of the sample in percent units.

14.4. Determination of MnO, CaO, and MgO Content of Concrete from a Single Weighing In

Principle of the Determination _____

The concrete sample is dissolved, the solution neutralized with ammonium hydroxide, and the metal hydroxides are precipitated.

Without filtering the precipitate, the manganous ions are oxidized with potassium permanganate to manganic ions, the calcium ions are precipitated with oxalate, and the magnesium ions with disodium hydrogen phosphate. In all three cases, the changes in temperature — proportional to the concentration of the component to be determined — are measured.

Apparatus

Directhermom apparatus
Plastic beakers, 300 cm³
Immersion pipettes
Magnetic stirrer rod
Beckman thermometer
Burettes, 25 cm³
Graduated cylinder, 20 cm³
Pipette, 10 cm³

Chemicals

Acid mixture for dissolution: 700 cm^3 water, 100 cm^3 nitric acid solution (sp.gr. 1.4) and 200 cm^3 hydrochloric acid solution (sp.gr. 1.19)

Concentrated ammonium hydroxide solution

Ammonium chloride solution, 20%

Potassium permanganate solution, saturated

Potassium oxalate solution (300 g potassium oxalate in 1000 cm^3)

Disodium hydrogen phosphate solution, saturated at room temperature and whose every 1000 cm^3 is diluted with 100 cm^3 water

Procedure

1.000 g of the finely powdered concrete sample is poured into a dry 600-cm^3 beaker, suspended in 30 cm^3 water with vigorous stirring; 50 cm^3 of the acid mixture is added and dissolved with warming. The mixture is kept boiling for 1 min, cooled, 50 cm^3 20% NH$_4$Cl solution is added to it, and the mixture is neutralized until incipient precipitation of the metal hydroxides. An excess of 5 cm^3 NH$_4$OH is added to the neutral solution. The solution is cooled and its volume is made up to 200 cm^3 in a graduated cylinder. The temperature of the solution is adjusted to room temperature and it is poured into the plastic beaker of the Directhermom apparatus. The beaker is placed into the Dewar flask. The immersion pipettes are filled with the following solutions: the 2-cm^3 pipette with KMnO$_4$ solution, the 10-cm^3 pipette with (COO)$_2$K$_2$ solution, and the 8-cm^3 immersion pipette with Na$_2$HPO$_4$ solution. Stirring is started and the pipettes and the thermistor are immersed into the solution. After a waiting period of 3 min (to attain temperature equalization) the shunt serving for the measurement of the percent MnO content is switched into circuit and the galvanometer is switched to "THERM." position and zeroed. The KMnO$_4$ solution is brought into the sample solution and the maximum galvanometer deflection is read. The shunt serving for CaO determination is substituted for the previous one, the galvanometer is repeatedly zeroed, and the (COO)$_2$K$_2$ reagent is discharged into the solution. Finally the shunt serving for MgO determination is switched on and after zeroing the galvanometer the Na$_2$HPO$_4$ solution is brought into the sample and the maximum deflection of the galvanometer is read.

Evaluation

Calculate the MnO, CaO, and MgO content of the concrete in percent units.

14.5.   References

Brown, M. E., *Introduction to Thermal Analysis,* Chapman & Hall, London, 1988.

Hemminger, W. F. and Cammenga, H. K., *Methoden der thermischen Analyse,* Springer-Verlag, Berlin, 1989.

Sesták, J., Thermophysical properties of solids. Their measurements and theoretical thermal analysis, in *Wilson and Wilson's Comprehensive Analytical Chemistry,* Vol. 12D, Svehla, G., Ed. Elsevier, Amsterdam, 1984.

Wendlandt, W. W., *Thermal Analysis,* 3rd ed., John Wiley & Sons, New York, 1986.

Chapter 15

Differential Scanning Calorimetry (DSC)

Principle of the Technique

Differential scanning calorimetry (DSC) has evolved from the technique of differential thermal analysis (DTA). There was no fundamental change in the measuring principle, only the position of the thermocouple sensors and controlling the transport of heat to the sample — keeping it at a nearly constant value — was improved by careful arrangement of the parts of the apparatus. A number of special solutions were applied in instrument design, which, however, will not be discussed here in detail.

The instruments used for DSC measurements can be divided into two groups, according to their basic principle of operation.

1. The temperatures of the sample and of the reference material are kept at an identical value, while the temperature of the reference material is increased according to a program. The recorder records the increase in electric power necessary to attain this identity of temperatures, as a function of time. The integral of the curve recorded in the course of a transition gives the heat of transition. An apparatus working along this principle is manufactured by the firm Perkin-Elmer.

2. The difference in the temperatures of the sample and the reference material are recorded (DTA principle) while the heat transport is controlled. Instruments of this type are manufactured by Du Pont (U.S.), Mettler (Switzerland), and Linseis (BRD).

In the case of instruments of the latter type, the sensing thermocouples are placed outside the sample holder, but in a close thermal contact with it. At the same time, better heat currents (thermal fluxes, cal/s) from the heating element to the reference material and the sample are realized by the application of a plate made of a good heat conductor.

The schematic diagram of the apparatus is presented in Figure 15.1.

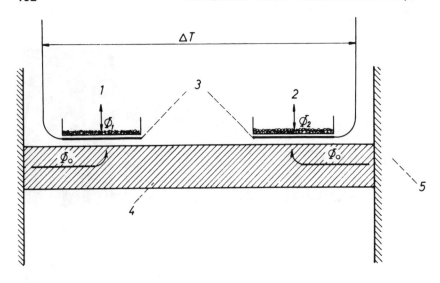

FIGURE 15.1

Schematic representation of a DSC apparatus. 1: Sample, 2: reference material, 3: temperature sensors, 4: thermal conductor, and 5: heating element.

It is apparent from the figure that, if there is no change going on in the sample in consequence of the increase in temperature, $\Phi_1 = 0$ and $\Delta T = 0$. As soon as a chemical or physical change starts in the sample, $\Phi_1 \neq 0$ and the following equation holds:

$$\Phi_1 = K_1 \Delta T \qquad K_1 = \text{constant} \qquad [\text{cal/s}^\circ\text{C}]$$

The thermal effect of the process can be determined by integration of the equation with respect to time:

$$\int_{t_1}^{t_2} \Phi_1 = K_1 \int_{t_1}^{t_2} \Delta T \, dt$$

where t_1 and t_2 are the times of starting and finishing the process, respectively.

It is apparent that

$$\int_{t_1}^{t_2} \Phi_1 = \Delta h$$

where

$$\Delta h = \Delta H \frac{\text{sample weighed in}}{\text{molecular weight}}$$

and accordingly

$$\Delta h = K_1 \int_{t_1}^{t_2} \Delta T \, dt \qquad \text{where } K_1 = \text{constant} \qquad [\text{cal/s}^\circ\text{C}]$$

For carrying out the measurement in practice, the following further considerations are necessary, taking the transformations of designations into consideration.

The thermocouple sensor supplies a voltage V proportional to the temperature difference:

$$\Delta T = k_1 \, \Delta V$$

and hence

$$\Delta h = K_2 \int_{t_1}^{t_2} \Delta V \, dt$$

where

$$K_2 = K_1 \cdot k_1 \qquad [cal/s°C]$$

The recorder records the voltage of the thermocouple and its deflection is proportional to the voltage:

$$\Delta V = k_2 \, \Delta m$$

and accordingly

$$\Delta k = K_3 \int_{t_1}^{t_2} \Delta m \, dt$$

where

$$K_3 = K_1 \, k_1 \, k_2 \qquad \left[\frac{cal}{s \cdot cm} \right]$$

The recorder plots the curve vs. time, and the chart movement (measured in terms of its length) is proportional to time:

$$dt = k_3 \, dl$$

and

$$\Delta h = K_4 \int_{l_1}^{l_2} \Delta m \, dl$$

where $K_4 = K_1 \cdot k_1 \cdot k_2 \cdot k_3 \left[\frac{cal}{cm^2} \right]$ is the so-called calorimetric constant, a figure characteristic of the sensitivity of the apparatus.

The integral $\int_{l_1}^{l_2} \Delta m \, dl$ is the surface area under the curve.

On the basis of the aforementioned we may write the equation

$$\Delta h = K_4 \cdot A$$

where A is the surface area under the curve which can be determined, for example, by graphic integration.

Calibration of the apparatus consists of the determination of the calorimetric constant K_4. This is carried out by means of a known transformation, generally by making a record with a standard material of known heat of fusion. From the amount weighed in and from the heat of fusion, Δh can be calculated; the surface area under the curve (A) can be determined on the basis of the record and hence the value of the calorimetric constant K_4 can be determined.

Very small thermal effects of the order of 0.1 mcal can be determined by means of DSC. Generally an accuracy of 2% can be attained, but in high-precision experimental series the error can be decreased below 1%.

The precision of DSC, well above that of DTA, enables the measurement of the temperature dependence of specific heats. The basis of such measurements is that, depending on the specific heat and the amount of sample (i.e., the heat capacity thereof) the baseline deviates from that recorded with an empty sample holder. The DSC technique is, as compared with the conventional specific heat measuring methods, advantageous because it is considerably more convenient and faster.

Frequently it is also possible to carry out kinetic calculations on the basis of the adequate mathematical treatment of the curve shape.

The technique is also widely applied for checking the purity of substances.

15.1.   Investigation of a Liquid-Crystalline (Mesomorphous) Transition

Principle of the Determination

Solid crystalline substances which accordingly show anisotropy in their properties are transformed to an isotropic liquid at their melting point. This involves that a material of a high degree of orientation is transformed into a one of low orientation:

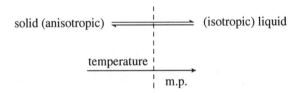

In the case of a number of organic compounds, between these two degrees of orientation there exits a third one, the so-called mesomorphous or liquid-crystalline state. In the case of such substances, the isotropic liquid state is reached from the anisotropic solid state through two points of transition:

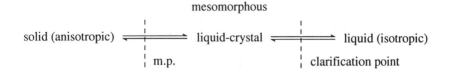

It is characteristic of a substance in the mesomorphous state that in some of its properties (e.g., mechanical) it behaves as a liquid, whereas in other (e.g., optical and electrical) ones it shows anisotropy.

By means of the differential scanning calorimeter it is possible to determine not only the characteristic temperatures of these transitions but, in spite of the low value of the heat requirement of mesomorphous-isotropic transitions, the latter can also be determined. By recording a cooling curve and the curve of repeated heating, even the monotropic and enantiotropic nature of these transitions can be observed.

Apparatus

DSC apparatus with temperature programmer
Preamplifier
X-Y recorder
Microbalance

Chemicals

Cholesterol capronate, 2 to 5 mg

Procedure

2 to 3 mg cholesterol capronate is weighed into the aluminum sample holder plate; the latter is put into the measuring cell and the cell is closed.

The temperature programmer is set to a heating rate of 5°C/min. The Dewar flask of the reference thermocouple is filled with ice.

A sensitivity of 3 µV/cm is adjusted on the Y channel of the recorder and that of 10°C/cm on the X channel.

Adjust the zero point of the temperature scale and start the temperature program and recording.

Simultaneously with this measurement observe the behavior of cholesterol capronate upon heating-cooling and repeated heating under a microscope equipped with a hot stage.

Make notes on the observations and determine the temperature values corresponding to the transformations.

Evaluation

Determine the individual transition temperatures from the DSC curves. On the basis of the DSC curves and the microscopic observations supplement the following transition scheme, showing the directions of the transitions with arrows and noting the corresponding transition temperatures:

S_1 (solid phase$_1$) L (liquid phase)
S_2 (solid phase$_2$) M (mesomorphous phase)

Decide whether the transformations are monotropic or enantiotropic.

15.2. Quantitative Calibration at One Temperature and Determination of the Heat of Transition of the β – α Crystal Modification of Potassium Nitrate

Principle of the Determination

On the basis of the principles described in the introduction, the differential scanning calorimeter also enables the determination of heats of transition.

First of all, the apparatus has to be calibrated, that is, the value of the constant K_4 has to be established. A standard substance should be chosen whose transition temperature, at which calibration is carried out, is near to the temperature of the transition to be measured. In this manner, the error brought about by the temperature dependence of the calorimetric constant can be decreased. The temperature pertaining to the β – α transition of potassium nitrate is, on the basis of data taken from the literature, 127.7°C and the heat of transition is 12.05 cal/g. Calibration against metallic indium as a standard seems to be adequate:

Melting point   156.6°C
Heat of fusion   6.79 cal/g

Weigh the potassium nitrate and the indium samples into the same sample holder. In order to determine the transition temperature, in the course of the first heating cycle record the thermogram as plotted against temperature. After cooling, in the second heating cycle, a recording is to be made as plotted against time, in order to determine the heat of transition.

Apparatus

DSC apparatus
Temperature programmer
Preamplifier
X-Y recorder
Microbalance

Chemicals

Potassium nitrate sample, 2 to 5 mg
High-purity indium standard, 2 to 5 mg

Procedure

Identical to that described in Exercise 15.1 with the only difference being in the second heating cycle, heating is made as plotted against time.

Evaluation

Determine the transition temperature of the $\beta - \alpha$ transition of potassium nitrate. Explain the origin of the deviation from data published in the literature.

Determine the calorimetric constant of the apparatus (K_4) by integration of the fusion curve of indium (surface area under the curve) and from the amount of indium weighed in.

Determine the heat of $\beta - \alpha$ transition of potassium nitrate from the integral of the transition curve, the calorimetric constant, and the amount weighed in.

Explain the deviation from the data given in the literature.

15.3. Checking of Purity and Determination of Melting Point

Principle of the Determination

The basis of the determination of purity by DSC is the known connection between the melting point depression of a material and the mole fraction of the impurity. In the case of low impurity concentrations, the following equation holds:

$$\Delta T = T_{mp} - T = \frac{RT_{mp}^2}{\Delta H_{sl}} X \qquad (1)$$

where T_{mp} is the melting point of the pure substance, T is the melting point of the sample, ΔH_{sl} is the heat of fusion of the pure substance, R is the gas constant, 1.987 cal/mol°C, and X is the mole fraction of the impurity.

According to the equation, the impurity (X) can be determined, if the melting point depression (ΔT), the melting point (T_{mp}), and the heat of fusion (ΔH_{sl}) are known. However, difficulties are encountered in practical measurements since in organic chemical practice generally neither the melting point of the pure substance (T_{mp}) nor the value of ΔH_{sl} are known to adequate accuracy. In spite of this fact, purity checking by means of the differential scanning calorimeter is widely used in practice, because in most cases a satisfactory accuracy can be attained by a single record.

Let us designate the cryoscopic constant in Equation 1 by K; in this case we may write

$$T = T_{mp} - KX$$

where

$$K = \frac{RT_{mp}^2}{\Delta H_{sl}} \tag{2}$$

The actual melting point T of the sample can be measured, but T_{mp}, ΔH_{sl}, and X are unknown.

If a fraction of the weighed sample, i.e., an r-th fraction of the weighing in is made to melt, the impurity will be enriched in this molten fraction and its mole fraction will increase to X/r. Consequently, the melting point, according to the previous equation, will be the following:

$$T = T_{mp} - \frac{KX}{r} \tag{3}$$

where

$$r = \text{melt fraction} \frac{\text{melt [mg]}}{\text{weighed in [mg]}}$$

Knowing the melting point (Tr) values pertaining to the melt fraction (r) values, and plotting the latter against 1/r, a straight line is obtained, whose point of intersection with the ordinate gives the T_{mp} value without the necessity of having the pure preparate. If the heat of fusion has been determined by another DSC measurement, the mole fraction of the impurity can be calculated from the slope (s) of the straight line:

$$s = K \cdot X = \frac{RT_{mp}^2}{\Delta H_{sl}} X \tag{4}$$

(see Exercise 15.1)

Accordingly, the heat of fusion of the pure material is substituted by that of the sample.

The only question that remains now is the following: how is it possible to obtain the function

$$T_4 = \frac{1}{r}$$

from the DSC record?

Figure 15.2 shows a typical DSC curve that can be used for purity testing. In this case the DSC curve or rather the ΔT axis is strongly expanded (the sensitivity of the XY recorder along the X axis was high).

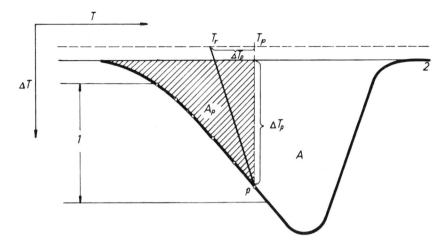

FIGURE 15.2
DSC curve. 1: Range of evaluation, r = 10 to 50%; 2: baseline of apparatus without sample.

The evaluation is carried out on the downward branch of the DSC curve in the

$$0.1 < \frac{A_p}{A} < 0.5$$

range at four to five points.

During fusion, the amount of material molten to the point P is proportional to the amount of heat input and that, in turn, to the area under that point, i.e., to A_p, and consequently the melt fraction can be calculated by the following equation:

$$r_p = \frac{A_p}{A} \tag{5}$$

where r_p is the melt fraction in point P, A_p is the area under the curve until point P, and A is the total area under the curve.

The temperature is recorded by a thermocouple placed on the reference side; this follows the temperature program and the value read in point P is T_p.

However, the temperature (T_r) of the sample of the melt fraction r_p is lower than that value by ΔT_p; the difference in the sample and reference temperatures and consequently T_r can be calculated from T_p by a simple connection:

$$T_r = T_p - \Delta T_p \tag{6}$$

The corresponding $T_r - 1/r$ value pairs are plotted in a diagram. Such a diagram is shown in Figure 15.3.

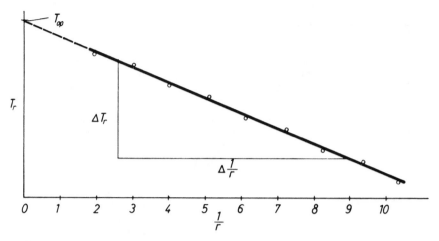

FIGURE 15.3

T_r vs. $\log \dfrac{[A]^-}{[HA]}$ diagram.

The diagram enables one to read the T_{mp} value and to calculate the slope:

$$s = \frac{\Delta T_r}{\Delta \dfrac{1}{r}} \tag{7}$$

and accordingly the degree of purity or the impurity content of the material can be calculated from Equation 4.

Purity checking by means of the DSC technique has a number of advantages compared to other methods: its simplicity, speed, and general applicability (since it is not necessary to know the quality of the impurity). Furthermore, the examination yields two more thermodynamic data: T_{mp} and ΔH_{sl}.

It is an advantage of the technique, which, however, can be regarded a drawback in some cases that adequate accuracy — about 1 to 2%, as calculated in terms of the mole fraction of the impurity — can be attained in the case of high-purity substances of a purity higher than 98 to 99%, whereas, in the case of purities of the order of 95%, the error in the mole fraction of the impurity may be as high as 20 to 25 relative percent.

The theoretical calculations are valid only for very highly diluted solutions or, in this case, melts.

The range of the applicability of the technique is narrowed by the following conditions of the calculations:

1. The components (sample and impurity) produce a phase diagram containing a eutectic composition; they do not form a mixed crystal.
2. The impurity or impurities form an ideal solution with the melt of the sample.

3. The heat of fusion is independent of temperature and the concentration of the impurity. (It is the heat of fusion of the sample already containing the impurity that is measured and, of course, at a temperature lower than T_{mp} by the value of the freezing point depression.)

Apparatus

> DSC apparatus
> Temperature programmer
> Preamplifier
> XY recorder
> Microbalance

Chemicals

> Phenacetine, containing 3 to 5 mg of p-amino benzoic acid or benzamide as impurity

Procedure

In the course of the first heating cycle, the heat of fusion of the sample is determined as described in connection with Exercise 15.2. In the second cycle, heating is made plotted against temperature and in the vicinity of the melting point the sensitivity of the recorder in the direction X is increased to a value of 2.5°C/cm.

Evaluation

Calculate the heat of fusion in calories per mole from the first record.

Determine the corresponding T_r and $1/r$ values at five points from the second record and draw the T_r vs. $1/r$ diagram.

Carry out the calculations also with the T_{mp} and H_{sl} values taken from the literature and compare the mole fraction values obtained in the two manners.

Explain the difference between the measured and the given impurity concentration values.

Part IV

Analysis of Organic Functional Groups

Chapter 16

Determination of Organic Functional Groups

16.1. Determination of Sorbitol and Glycerol When Present Together by Oxidation With Periodate According to Maros and Schulek

The oxidation of α-glycols, polyhydroxy compounds with periodate in aqueous solution, is an unequivocal, fast, and selective reaction. The reaction is interpreted by assuming the cleavage of the bond between the carbon atoms bearing the two vicinal periodate-active groups and attaching a hydroxyl group to both of them. Polyoxy compounds are oxidized by periodate in a multistep reaction. The compounds formed in the first step are broken up into smaller fragments in subsequent steps, until there is no more vicinal periodate-active group in the compound.

The general equation of the oxidation is

$$\begin{array}{l} CH_2OH \\ | \\ (CHOH)_n + (n+1) IO_4^- = 2CH_2O + n\,HCOOH + (n+1) IO_3^- + H_2O \\ | \\ CH_2OH \end{array}$$

The reaction is fast and unequivocal in acidic, neutral, or weakly alkaline medium, at room temperature.

By measuring the excess of periodate and the stable products of oxidation (formaldehyde, formic acid), the weight ratio of two known polyalcohols in a mixture can be determined.

Measurement of Excess Periodate

Principle of the Determination
In weakly alkaline solution periodate oxidizes iodide but iodate does not. In a solution containing sodium hydrogen carbonate (pH = 8) periodate is reduced using iodide:

$$IO_4^- + 2 I^- + 2H^+ = I_2 + IO_3^- + H_2O$$

and the iodine formed can be titrated with a sodium arsenite standard solution.

Apparatus

Automatic burettes
Conical flasks with ground-glass stopper, 250 cm^3
Graduated cylinder, 100 cm^3
Pipettes with two marks, 10 cm^3
Burette, 12 cm^3

Chemicals

0.1 N sodium arsenite solution
0.15 M periodic acid solution
Starch indicator
Sodium hydrogen carbonate, solid
Potassium iodide, solid

Procedure
Transfer a 10.00-cm^3 aliquot of the unknown containing about 1 mequiv of sorbitol and glycerol to a conical flask with ground-glass stopper and add 10.00 cm^3 of 0.15 M periodic acid solution. After 30 min of waiting, dilute to 50 cm^3, add about 1 g of sodium hydrogen carbonate, and after the evolution of carbon dioxide shake the contents of the flask vigorously (opening the flask from time to time). Now add 0.5 to 1 g of potassium iodide, and titrate the iodine formed with 0.1 N sodium arsenite standard solution, in the presence of eight to ten drops of starch solution. Carry out a blank test and use the difference of the two titrations when calculating the result: 1 cm^3 of 0.1 N periodic acid measures 1.822 mg of sorbitol or 2.302 of glycerol.

Measurement of Formaldehyde Using the Hydrogen Sulfite-Cyanide Method

Principle of the Determination _____

The polyoxy compound is oxidized with periodate and after waiting the excess of periodate and the iodate is reduced to iodide with sodium sulfite. Formaldehyde formed in the oxidation is transformed into formaldehyde hydrogen sulfite by adding sulfite in excess; the excess of sulfite is oxidized with iodine. Then the solution is alkalized and the aldehyde hydrogen sulfite is decomposed with cyanide, using pentane for excluding oxygen:

$$CH_2OHSO_3^- + CN^- = CH_2OHCN + SO_3^{2-}$$

After acidifying, the sulfurous acid formed in an amount equivalent to that of the aldehyde originally present is titrated with iodine standard solution. The amount of standard solution consumed is proportional to that of formaldehyde or polyalcohol. The equivalent weight of polyalcohols is equal to one fourth of the molecular weight.

Apparatus

Pipette with two marks, 10 cm^3
Graduated pipettes, 12 cm^3
Conical flasks with ground-glass stoppers, 100 cm^3
Burette, 12 cm^3

Chemicals

0.1 N iodine solution
5% iodine solution
0.15 M periodic acid
1.5 M sodium sulfite solution
10% acetic acid solution
20% hydrochloric acid
20% sodium hydroxide solution
Methyl red indicator
Starch indicator
Potassium cyanide, solid
Pentane

Procedure

Transfer a 10.00-cm^3 aliquot of the stock solution containing about 1 mequiv of sorbitol and glycerol to a 100-cm^3 conical flask with ground-glass stopper and add 10.00 cm^3 of 0.15 M periodic acid. After 30 min waiting and sodium sulfite solution dropwise until the color of iodine disappears and add 0.5 to 1 cm^3 in excess. Make the solution slightly alkaline with sodium hydroxide in the presence of methyl red, then neutralize with acetic acid and add 1 cm^3 in excess.

Cool the warm solution, shake the flask, opening the stopper from time to time. Pour 5 cm^3 of pentane onto the surface of the solution, and after 20 to 30 min waiting, oxidize the bulk of excess sulfite with 5% iodine solution, in the presence of ten drops of starch solution. Oxidize the remainders of excess sulfite with 0.1 N iodine solution added from a burette. After the appearance of the blue iodine-starch complex shake the contents of the flask vigorously, opening it from time to time and, if necessary, adjust the color to light blue with one or two drops of iodine solution.

After spreading of the pentane, add 3 cm^3 of 20% sodium hydroxide and 0.2 to 0.3 g of solid potassium cyanide, and after closing the flask mix cautiously. After 2 to 3 min waiting add one drop of methyl red and neutralize with hydrochloric acid and add 2 cm^3 in excess. Then titrate with 0.1 N iodine solution, until the appearance of the blue color, shaking cautiously.

Closing the flask, shake again, and if necessary add one or two drops of the iodine solution to obtain a stable blue color.

1 cm^3 of 0.1 N iodine is equivalent to 4.554 mg sorbitol and 2.302 mg glycerol.

Evaluation

Give the result:

Sorbitol $=$ mg/100 cm^3

Glycerol $=$ mg/100 cm^3

Note: HCN is poisonous! Work very cautiously with the acidified solution of cyanide!

16.2.   Determination of Alkoxy Groups

Principle of the Determination _____

In most cases methoxy and ethoxy groups are to be determined.

Methyl and ethyl groups bound to the rest of the molecule through oxygen, or in some cases through sulfur, can be transformed into methyl or ethyl iodide, respectively, upon boiling with hydrogen iodide:

$$R{-}OCH_3 + HI = R{-}OH + CH_3I \qquad \text{(boiling point : 42.5°C)}$$

$$R{-}OC_2H_5 + HI = R{-}OH + C_2H_5I \qquad \text{(boiling point : 72.2°C)}$$

The product can be distilled off. During distillation carbon dioxide is passed through the apparatus.

Phenol and propionic anhydride is used as solvent of the substance to be investigated, and concentrated hydrogen iodide (density: 1.7 g/cm^3) is added. Besides the alkyl iodide, hydrogen iodide, iodine, and from sulfur-containing compounds also hydrogen sulfide distills off. They have to be removed from the alkyl iodide vapor. Accordingly the vapor is passed through a washing liquid containing cadmium sulfate and sodium sulfate.

On passing alkyl iodide vapors through a solution of bromine in glacial acetic acid, methyl bromide and bromine-iodide is formed:

$$CH_3I + Br_2 = CH_3Br + BrI$$

By the action of excess bromine, bromine-iodide decomposes, and iodine is oxidized to hydrogen iodate:

$$BrI + 2Br_2 + 3H_2O = HIO_3 + 5HBr$$

By reducing the excess of bromine with formic acid, and then after acidifying with sulfuric acid, and adding potassium iodide, iodine is formed:

$$Br_2 + HCOOH = CO_2 + 2HBr$$

$$IO_3^- + 5I^- + 6H^+ = 3I_2 + 3H_2O$$

The iodine can be titrated with a sodium thiosulfate standard solution, using starch as indicator. The determination is very sensitive, since six equivalents of iodine correspond to one alkoxy group.

Apparatus

CO_2 gas cylinder
Distilling apparatus
Pipettes, 2 cm^3, 2 cm^3 graduated
Burette
Titration flask

Chemicals

Bromine in glacial acetic acid
Acetic anhydride
Phenol
Hydrogen iodide, concentrated (sp.gr.1.7)
Concentrated formic acid
1:3 diluted sulfuric acid
Potassium iodide, solid
0.02 N sodium thiosulfate standard solution
Starch
Trichloroethylene

Procedure

Assemble the distilling apparatus according to Figure 16.1. Fill the washing vessel (c) with the washing solution to half of its height, then the two absorption vessels (d_1, d_2) with bromine solution in glacial acetic acid, filling the first vessel to two thirds, the second one to one third of its height. Connect the condenser (b) to the tap, fasten all the joints with rubber rings.

Then place the methoxy-containing sample into the spherical flask (a) of the apparatus. Add 1.5 cm^3 acetic anhydride and 0.1 to 0.2 g of phenol. Heat cautiously with a burner until dissolution, then cool the flask. Connect the side tube for gas introduction with the CO_2 gas cylinder, then add 2 cm^3 of hydrogen iodide using a pipette under cooling, and fix the flask immediately to the condenser and fasten the joint with a rubber ring. Start the carbon dioxide stream, at a rate of 1 bubble per second. Start heating the flask and keep the liquid in quiet boiling for 3/4 hour. Then disassemble the apparatus, wash and dry the flask. The washing solution can be used in other determinations; thus, it can be left in the vessel.

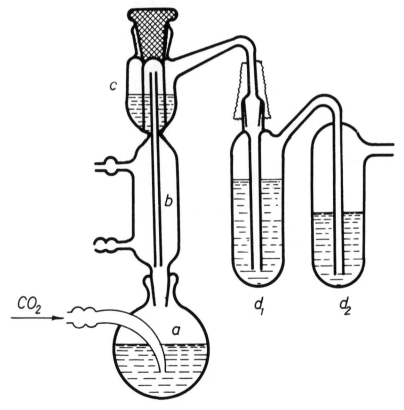

FIGURE 16.1
Apparatus for the determination of alkoxy groups.

Pour the contents of the two absorption flasks into a titration flask, using distilled water for rinsing the vessels. Add some drops of concentrated formic acid until the color of bromine disappears. Then acidify the solution with 5 cm^3 of 1:3 diluted sulfuric acid and add about 0.5 g of potassium iodide. Titrate the iodine formed with 0.02 N sodium thiosulfate standard solution, using starch as indicator.

1 cm^3 of 0.02 N sodium thiosulfate solution corresponds to 0.1034 mg methoxy and 0.1502 mg ethoxy group.

The determination of ethoxy group is similar to that of methoxy, but trichloroethylene vapor (boiling point 96°C) is to be circulated in the condenser to prevent condensation of ethyl iodide having a higher boiling point.

Evaluation

Calculate the methoxy content of the substance studied in milligrams.

16.3. Determination of the Halogen Content of Organic Compounds Using the Oxygen-Flask Method and Potentiometric Titration

Principle of the Determination

By burning the halogen-containing organic sample in an oxygen atmosphere, the halogens are transformed into the corresponding hydrogen halides. The latter can be determined by potentiometric titration with silver nitrate after absorption in ammonium hydroxide.

In the determination of chloride, a silver electrode, or a chloride ion-selective electrode, for that of iodide an iodide ion-selective electrode can be applied. In all three cases a calomel electrode can be used as a reference electrode.

Apparatus

> Oxygen flask
> Oxygen gas cylinder with pressure gauge
> Bunsen burner
> Protecting cylinder made of wire net (the flask is put within it during burning of the organic substance)
> Precision pH meter
> Silver electrode or appropriate ion-selective electrode
> Saturated calomel electrode
> Magnetic stirrer

Beaker, 100 cm^3
Pipette, 10 cm^3, graduated
Burette, 20 cm^3
KNO$_3$-agar bridge

Chemicals

0.01 N silver nitrate standard solution
Phenolphthalein indicator
1:1 diluted acetic acid
1:1 diluted ammonium hydroxide
Potassium chloride, solid

Procedure

Weigh 15 to 25 mg of the compound to be studied by back weighing on an analytical balance, onto an L-shaped filter paper (Figure 16.2). Fold the paper in a way that the thinner part of the L can be used as an igniting strip. The number of the sample can be marked on this strip.

Ignite the platinum wire net or spiral sealed into the ground-glass stopper of the flask (Figure 16.3) in the upper part of a flame (do not ignite in carbonizing flame or in the reducing part of the flame). After cooling, fasten the filter paper containing the sample into it.

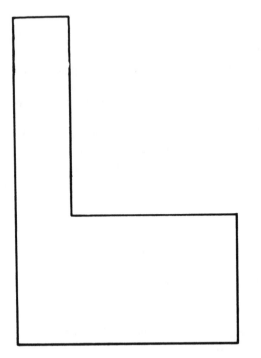

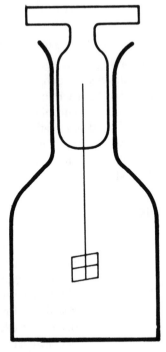

FIGURE 16.2
L-shaped filter-paper for burning.

FIGURE 16.3
Oxygen flask.

Pour 10 cm^3 of distilled water and 5 cm^3 of 1:1 ammonium hydroxide into the flask. Introduce oxygen into the bottom of the flask, right above the absorption liquid for some seconds. After stopping the gas flow, insert the flask within the wire-net cylinder. Ignite the end of the filter paper using a burner and insert the stopper immediately into the oxygen flask. Pour some distilled water around the stopper, into the collar of the flask. Thus any bubbling of the products of oxidation through the possibly loose stopper can be perceived. This can be prevented by wetting the ground glass previously with some distilled water and holding the stopper pressed downward during the whole course of ignition.

The hydrogen halide formed reacts with ammonia in the gaseous phase, forming ammonium halide, which has to be absorbed in the liquid by shaking the flask.

Before opening the flask, pour some distilled water into the collar of the flask. Put the flask on the table, press downward with one hand, and cautiously raise the stopper with the other, applying twisting (there is a slight vacuum in the flask due to the absorption of the oxidation products). Rinse the stopper with some water into the flask. Transfer the solution into a 100-cm^3 beaker rinsing the flask with small portions of water.

Place a stirring rod into the solution, place the beaker into a magnetic stirrer, and start the stirring. Add two or three drops of phenolphthalein and 1:1 diluted acetic acid dropwise from a graduated pipette to neutralize. If the phenolphthalein remains colorless in the original solution, alkalize the solution with 1:1 ammonium hydroxide, then do as given above.

Dip the appropriate indicator electrode into the solution to be investigated and the reference electrode into KCl solution. Connect the electrodes with the pH meter and the two solutions with each other by means of the salt bridge. (The ends of the salt bridge must not hold air bubbles.)

Set the instrument into millivolt position and find the actual range. (Calibration is necessary only if direct activity measurement is made, but not in recording the titration curve.)

Add the titrant, 0.01 N silver nitrate standard solution, in 0.5-cm^3 portions at the beginning and in 0.1-cm^3 portions when getting closer to the equivalence point. Add some 0.5-cm^3 portions to reach 2 to 3 cm^3 excess after the equivalence point. Read the potential after each addition.

Collect the data in a table.

Evaluation

Plot the potentiometric titration curve, find the equivalence point, and calculate the halide content of the sample in percent units.

Part V

Separation Techniques

Chapter

Ion Exchangers and Their Analytical Applications

Principle of the Technique

Ion exchangers are water-insoluble, mostly solid substances which, when contacted with aqueous electrolyte solutions, take up positively or negatively charged ions from the latter and simultaneously release an equivalent amount of other, identically charged ions into the solution. According to the sign of the ions exchanged, cationic and anionic ion exchangers can be distinguished. The natural-base (cellulose, dextrane) and plastic resin-base (mostly divinyl-benzene and styrene copolymers) ion exchangers are the most widely used. These contain various functional groups. The functional groups of cation exchangers are, for example, SO_3H (strongly acidic) or $COOH$ (weakly acidic) groups. These groups bind the cations and release protons into the solution. The functional groups of anion exchangers are, for example, NR_2 (weakly basic) or NR_3 (strongly basic). These groups bind anions and release OH ions into the solution.

Strongly acidic cation exchanger resins are apt to exchange ions in a wide pH range (1 to 11 pH) and are capable to bind metal ions. Weakly acidic resins can be used in a narrower range (6 to 10 pH). The operation of anion exchangers is also dependent on pH.

Proper selection of the ion exchanger applied and of the medium enables a number of separation tasks to be carried out.

17.1. Preparation of an Ion-Exchange Column and Determination of the Salt-Decomposition Capacity

Principle of the Determination

If a large amount of sodium chloride solution is poured onto a strongly acidic cation-exchange column loaded with hydrogen ions, the effluent will contain, in addition to sodium chloride, hydrogen ions displaced from the column, which can be titrated with alkali standard solution.

The processes can be described by the following reaction equations:

$$R–H + NaCl = R–Na + HCl \qquad \text{(ion exchange)}$$

$$HCl + NaOH = NaCl + H_2O \qquad \text{(titration)}$$

The salt-decomposition capacity of the ion-exchange resin is defined as the amount of hydrogen ions referred to unit mass of the dry resin or unit volume of the swollen resin.

Apparatus

Ion-exchange column with stand
Burette
Titrating flask
Beaker
Graduated cylinder

Chemicals

1 *N* sodium hydroxide solution
2 *N* hydrochloric acid solution
1 *N* sodium chloride solution
1% aqueous methyl red indicator solution
1 *N* sodium hydroxide standard solution

Procedure

Prepare a column of the strongly acidic ion exchange resin in the hydrogen form.

The column shown in Figure 17.1 equipped with overflow pipe is easy to prepare and to handle, its advantage being that the resin is always covered with liquid; thus, it is always swollen, ready for use. Dry or slightly wet cation exchange resins, which are commercially available, are to be swollen for 1 day in distilled water before. Then, on the next day, pour it in about 4 N hydrochloric acid and allow to stand for 1 d. If the acid becomes colored [most cation exchangers contain remarkable amounts of iron(III)], repeat the treatment until the 4 N hydrochloric acid remains colorless. Then wash the resin with water, add 100 cm^3 of 1 N sodium hydroxide, and allow to stand for 2 h.

After discarding the alkali, wash the resin with water, then pour on 1 N hydrochloric acid, shake from time to time, and pour off the acid.

Assemble the ion-exchange column and clamp to the stand. Fill it with water and press some glass wool into the bottom of the column with a glass rod. Pour the resin into the column in small portions, taking care that no bubbles get in between the resin particles.

Pour on 100 cm^3 of 1 N sodium chloride on the column, collecting the effluent in a titration flask, then adjust the flow rate to 2 to 5 cm^3/min by means of the stopcock of the separating funnel.

After all the solution has been poured on, add one or two drops of methyl orange indicator to the solution collected in the titration flask and titrate with 1 N sodium hydroxide standard solution. (Potentiometry can also be used for end-point indication.)

Calculate the salt-decomposition capacity of the column in milliequivalent units. To calculate the capacity referred to unit column volume, calculate the volume of the resin column after measuring the diameter and length. Finally, wash the column with distilled water.

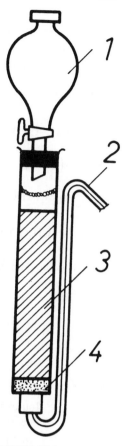

FIGURE 17.1
Ion-exchange column.
1: Separating funnel,
2: overflow pipe,
3: ion-exchange resin, and
4: glass wool.

Evaluation

The salt-decomposition capacity of the column is given in milliequivalents and the volume capacity in milliequivalents per cubic centimeters.

17.2.   Determination of Acetyl Groups

Principle of the Determination _____

The ester-containing acetyl group is hydrolyzed in alkaline medium. The alkali acetate formed is poured onto a strongly acidic cation-exchange column in the hydrogen form. By titrating the acetic acid in the effluent with alkali standard solution, the amount of acetyl group present in the ester can be calculated. The steps of the determination are

1. Hydrolysis:

$$CH_3\text{--}COOR + KOH \xrightarrow{\text{boiling}} CH_3COOK + R\text{--}OH$$

where R stands for an aliphatic chain

2. Ion exchange:

$$CH_3COOK + R'\text{--}H \longrightarrow CH_3\text{--}COOH + R'\text{--}K$$

where R′ is the equivalent amount of the resin

3. Titration:

$$CH_3\text{--}COOH + NaOH = CH_3\text{--}COONa + H_2O$$

Apparatus

> Ion-exchange resin column
> Separating funnel fixed to a stand
> Spherical flask and condenser with ground glass joint,
>   on a stand
> Bunsen burner
> Asbestos wire gauze
> Pipette, graduated
> Pipette, 5 cm^3
> Titration flask
> Funnel
> Graduated cylinders, 100 cm^3, 20 cm^3
> Burette

Chemicals

> 0.02 N carbonate-free sodium hydroxide standard solution
> 0.02 N oxalic acid solution (f = 1.000)
> 0.1% phenolphthalein indicator in alcohol
> 2 N hydrochloric acid
> 2 N potassium hydroxide in alcohol
> 1:1 mixture of methyl alcohol and water

Procedure

1. Pretreatment of the Ion-Exchange Resin

Pour 100 cm³ of 2 N hydrochloric acid on the ion-exchange column to regenerate the strongly acidic ion-exchange resin. Adjust the flow rate to 2 to 5 cm³/min by means of the stopcock of the separating funnel. Then wash the column acid-free with distilled water. Collect the effluent in a 100-cm³ graduated cylinder. After collecting 100 cm³ of the effluent, titrate it with 0.02 N sodium hydroxide standard solution.

Collect further 100-cm³ portions. Continue washing until the titrant solution required by two subsequent portions is the same (maximum 0.2 cm³). Note this consumption and treat it as a blank value. Check the blank value by titrating 100 cm³ of distilled water.

Then wash the acid-free column with 10 cm³ of 1:1 diluted methyl alcohol.

2. Hydrolysis

This operation can be done simultaneously with the regeneration of the ion-exchange resin.

Pour 5 cm³ of alcoholic 2 N potassium hydroxide solution into the spherical flask of the hydrolyzing apparatus using a graduated pipette. Take care that no alkali gets on the stopper.

Transfer a 5.00-cm³ aliquot of the unknown butyl acetate solution into the spherical flask with a pipette, put in a few glass beads to prevent retarded boiling, join the condenser, and start the cooling water. Heat the flask with a small flame through an asbestos wire gauze for 30 to 40 min after boiling has started. After hydrolysis, allow the reaction mixture to cool in the apparatus.

3. Ion Exchange

Put a titration flask under the overflow pipe of the ion-exchange column, and pour the contents of the spherical flask into the clean and empty separating funnel, taking care that no glass bead gets into the funnel. Open the stopcock. After all the solution has passed through, rinse the distilling flask with 10 cm³ 1:1 methyl alcohol and also pour this into the separating funnel. Repeat this rinsing twice, but wait in each case until the previous portion has completely passed through. Finally, wash the column with 50 cm³ distilled water poured in small portions and titrate the collected effluent. Make a blank test.

4. Titration

Add two or three drops of phenolphthalein to the content of the two titration flasks (sample and blank) and titrate with 0.02 N carbonate-free sodium hydroxide standard solution to pink. The color of the indicator should remain pink for about half a minute.

If the blank does not require standard solution in noticeable amount, the previous washing has been good. If the blank value is too high, wash the ion-exchange column further or add the requirement of the blank to that of the sample.

Standardize the sodium hydroxide solution by means of 0.02 N oxalic acid (f = 1.000), using phenolphthalein as indicator.

Subtract the proportional fraction of the blank value determined for 200 cm^3 of distilled water from the standard solution required by the sample.

Evaluation

Calculate the amount of butyl acetate in milligrams per cubic centimeters.

17.3. Determination of Phosphate Ions from a Solution of the Salt by Potentiometric Titration Following Ion Exchange

Principle of the Determination

The solution containing phosphate and other anions and various metal ions is poured on a strongly acidic cation-exchange column loaded with hydrogen ions. The phosphoric acid in the effluent is determined with alkali standard solution using visual or potentiometric end-point indication.

The determination can be described by the following reaction equations:

Ion exchange:

$$4R\text{--}H + K_3PO_4 + KCl = 4R\text{--}K + H_3PO_4 + HCl$$

Titration I:

$$H_3PO_4 + HCl + 2NaOH = NaH_2PO_4 + NaCl + 2H_2O$$

Titration II:

$$NaH_2PO_4 + NaOH = Na_2HPO_4 + H_2O$$

Apparatus

Volumetric flask
Pipette, 10 cm^3
Beaker or titration flask
Ion-exchange resin column

pH meter
Magnetic stirrer
Glass electrode
Saturated calomel electrode

Chemicals

0.1 N sodium hydroxide standard solution
0.1 N hydrochloric acid standard solution
2 N hydrochloric acid solution

Procedure

Regenerate the strongly acidic cation-exchange resin column and wash it acid-free with distilled water (see Section 17.1). Make up the unknown phosphate solution to the mark in a 100-cm^3 volumetric flask. Transfer 10.00 cm^3 of this solution to the separating funnel of the ion-exchanger device, and place a beaker under the overflow pipe. Adjust a flow rate not exceeding 5 cm^3/min. After the sample has passed through, wash the column with 4×10 cm^3 distilled water, waiting for each portion to pass through completely before pouring on the next one. Pour on further 50 cm^3 of distilled water and check that the effluent is acid-free, using methyl orange of indicator paper.

Titrate the collected effluent potentiometrically with 0.1 N NaOH standard solution, using a glass electrode as indicator and a saturated calomel electrode as reference electrode. Stir the solution during titration. Add the titrant in 1-cm^3 portions at the beginning and in 0.1-cm^3 portions close to the equivalence point of the titration, and read the pH after each addition. Complete the titration by adding 3 to 4 cm^3-s of the titrant in excess, in portions.

The PO_4^{3-} or P_2O_5 content of the solution can be calculated on the basis of the titration curve. Regenerate the ion-exchange column after the determination with 100 cm^3 of 2 N hydrochloric acid and distilled water.

Evaluation

Plot the potentiometric titration curve and calculate the PO_4^{3-}, and P_2O_5 content of the sample solution in milligrams per cubic centimeters.

17.4. Determination of Iron(III) and Zinc(II) Following Separation by Ion Exchange

Principle of the Determination _____

The separation is based on the difference in the stabilities of the negatively charged chloro-complexes of the two metals, that of the zinc chloro-complex being higher. The stability can be influenced by changing the chloride concentration in the solution. The condition for the separation of two ions is that the ratio of the distribution coefficients differs from 1

$$\frac{D_B}{D_A} = K_d \gtrless 1$$

The separation is effected using a strongly basic anion-exchange resin column in the chloride form. On pouring the solution containing zinc(II), iron(III), and chloride ions on the column, the zinc complex is bound by the resin whereas the iron(III) complex passes through. Zinc can be eluted after decomposition of the complex.

Apparatus

> Ion-exchange column
> Titration flask
> Beaker
> Microburette
> Pipette, 10 cm^3
> Graduated cylinder, 100 cm^3

Chemicals

> Eluting solution I: Dissolve 100 g sodium chloride in some distilled water, add 10 cm^3 concentrated hydrochloric acid and make up to 1 l in a volumetric flask
> Eluting solution II: Dissolve 20 g ammonium chloride in some distilled water, add 55 cm^3 of concentrated ammonium hydroxide and dilute to 1 l in a volumetric flask
> 0.05 M and 0.25 M EDTA standard solution
> Tirone indicator
> Eriochrome Black T indicator
> Glacial acetic acid
> 1:1 ammonium hydroxide
> 1 N hydrochloric acid

Procedure

Prepare a column of strongly basic anion-exchange resin, transform into the chloride form with about 50 cm^3 of 1 N hydrochloric acid, and wash it acid-free with distilled water, then pour onto the column 20 cm^3 of the eluting solution I (do not wash the column afterward!).

Make up the unknown solution to 100 cm^3 with eluting solution I in a volumetric flask. Transfer 10.00 cm^3 of this solution to the separating funnel above the column. Adjust a flow rate of 2 to 3 cm^3/min. Collect the effluent in a titration flask. After the solution has passed through, wash the column with 100 cm^3 eluting solution I, poured on in three equal portions. Put aside the collected effluent and place another flask below the overflow pipe.

Elute zinc(II) with 100 cm^3 of eluting solution II, poured on in three equal portions. By this solution positively charged zinc(II) amine complex ions are formed which are not bound by the ion-exchange column.

Determination of Iron(III)

Pour 2 to 3 cm^3 of glacial acetic acid into the titration flask containing iron(III) to adjust the pH, and add 1:1 ammonium hydroxide dropwise from a graduated pipette, under stirring until the appearance of the reddish brown color of the iron(III) acetate. Warm the solution to 50 to 60°C, add 2 to 3 cm^3 Tirone indicator solution and titrate with 0.05 *M* EDTA standard solution added from a microburette. (The color of the indicator changes to pale yellow from blueish green.)

Determination of Zinc(II)

The pH necessary in the titration is adjusted by the eluting solution II. Titrate in the presence of Eriochrome Black T as indicator with 0.025 *M* EDTA standard solution. (Color transition at the equivalence point: from violet to blue.)

After determination wash out the column with 50 cm^3 distilled water and regenerate with 50 cm^3 1 *N* hydrochloric acid and wash with distilled water.

Evaluation

Calculate the iron(III) and zinc(II) contained in the unknown solution in milligrams per cubic centimeters.

17.5. References

Inczédy, J., *Analytical Applications of Ion Exchangers*, Pergamon Press, Oxford, 1966.

Tarter, J. G., Ed., *Ion Chromatography*, Chromatographic Science Series, Vol. 37, Marcel Dekker, New York, 1987.

Chapter 18

Thin-Layer Chromatography

Principle of the Technique —————————————————————

Thin-layer chromatography is a variant of liquid chromatography. Its essence is that the stationary phase is spread on a relatively large, plane surface (glass, aluminum, or acetyl cellulose plate) in a thin layer of the thickness of 0.1 to 0.4 mm. The material of the stationary phase is generally silica gel (e.g., Kieselgur G), mixed with some gypsum in order to attain good adhesion to plate, or else aluminum oxide.

18.1. Identification of the Components of Mixtures of Phenol, Aldehyde, and Azo Compounds After Separation by Thin-Layer Chromatography. Semiquantitative Determination of Vanillin. Quantitative Determination of p-Amino Azobenzene

Apparatus

> Plastic sheets, 200 × 100 mm, 200 × 80 mm
> Pipettes or capillaries for spot application
> Separation chamber
> Spray apparatus for development
> Adsorbent: Kieselgel G, Merck

Development

Glass chambers with lids are used for development. Pour the developing solvent to form a 5- to 7-mm layer. To ensure saturation with the solvent, line the sides of the chamber with filter paper saturated with the solvent.

Place the coated sheet into the chamber and cover it with the lid. Allow to develop until the front of the solvent reaches the front distance previously marked and then take out the sheet. Draw the front line on the wet sheet and dry.

If the substances separated are colored, no visualization is necessary. In other cases the spots can be visualized by spraying with suitable reagent and/ or viewing under ultraviolet light.

Evaluation

1. Identification is made on the basis of the R_f values.

$$R_f = \frac{\text{distance of the center of the spot from the start point}}{\text{distance of the front of the solvent from the start point}}$$

2. Semiquantitative evaluation on the basis of the relationship between the size of spot and the amount of substance.

3. Quantitative determination can be done by a great variety of methods, either by measuring some properties of the spot which are proportional to the concentration or after removing the spot from the glass sheet, by a suitable analytical method.

Identification of the Components of Mixtures of Phenols or Aldehydes or Azo Compounds

Procedure

The composition of the mixtures are as follows:

Phenols	Naphthol
	Chloroglucinol
	Pyrogallol
	Gallic acid
	Resorcinol
Solvent	Ethyl alcohol
Developing agent	A mixture of dioxane, benzene, and acetic acid (25:90:5)
Visualizing	By spraying with a 20% solution of $SbCl_3$ in chloroform and drying in a drying oven

Aldehydes	Salicyl aldehyde, p-nitro salicyl aldehyde, vanillin
Solvent	Ethyl alcohol
Developing agent	Chloroform
Visualizing	By spraying with a 4% solution of 2,4-dinitro phenylhydrazine in 2 N hydrochloric acid
Azo compounds	Azobenzene p-Amino azobenzene Dimethylamino azobenzene p-Hydroxy azobenzene
Solvent	Ethyl alcohol
Developing agent	Benzene
Visualizing	By treating with vapors over cc HCl

A mixture of one of the compound types containing an unknown number of components is to be analyzed. (The pure components and their mixtures are available.)

Evaluation

The components found and their R_f values and those of the standards are given.

Semiquantitative Determination of Vanillin

Procedure

Apply 2, 4, 6, and 8 µl of the ethyl alcoholic vanillin solution of the concentration 20 mg/l with a micropipette onto four neighboring strips of a coated sheet, as well as 2, 4, 6, and 8 µl of the vanillin solution of unknown concentration prepared with ethanol. Take care that the spot sizes are identical as far as possible.

Use chloroform as a developing agent and a 0.4% solution of 2,4-dinitro phenylhydrazine in 2 N hydrochloric acid for visualization. After development and spraying with the reagent solution, calculate the areas of the spots. For elliptic spots the area, F, can be calculated as

$$F = r_1 \cdot r_2 \pi$$

where r_1 and r_2 are the radii of the ellipse.

Evaluation

Prepare a calibration graph (Figure 18.1), putting the square root of the spot area on the vertical and the logarithm of the amount of substance applied on the horizontal axis. In a certain concentration range, the following relationship exists:

$$\sqrt{F} = a \cdot \log M + b$$

where $\sqrt{F}$ is the square root of the spot area, log M is the logarithm of the amount of substance applied, and a and b are constants characteristic of the compound studied.

Calculate the values of a and b on the basis of the calibration graph.

Determine the spot areas for the unknown and the concentration of the unknown vanillin solution in milligrams per cubic centimeters, using the a and b values determined previously.

Quantitative Determination of p-Amino Azobenzene Using a Densitometer

Principle of the Determination _____

The components can be determined on the sheet by a densitometer, following separation.

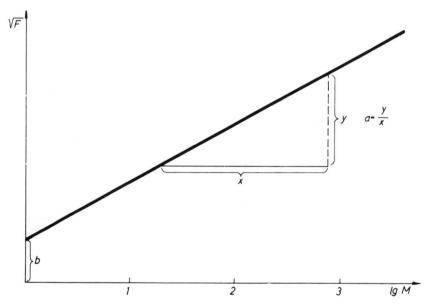

FIGURE 18.1
Calibration diagram to semiquantitative determination.

Procedure

Prepare the chosen sheets for development: divide them into two 4-cm strips longitudinally. Only two samples can be evaluated on a sheet by the densitometer (Figure 18.2).

Apply 2, 4, 6, and 8 µl of the 2 mg/cm³ p-amino-azobenzene stock solution to the four start points of a sheet, using a Desaga micropipette or a Hamilton syringe. After drying the spots, place the sheet into benzene developing agent. Run until the front reaches nearly half of the sheet, mark the solvent front, invert the sheet, and run the other side as well. After the solvent front has reached 8 to 10 cm above the start point, finish the development and mark the solvent front.

Prepare the thin layer chromatogram of the unknown sample similarly, using 2-, 4-, 8-, and 10-µl volumes. The chromatograms need not be visualized over concentrated hydrochloric acid.

Put the densitometer in operation according to the instructions for use and record the absorbance diagrams of the sheets.

The spot of a component appears as a peak on the curve (Figure 18.3).

The amount of substance present in the spot is proportional to the peak area. To quantitative evaluation the curve is to be integrated which is done by a mechanical integrator. The integral is recorded simultaneously (Figure 18.4).

The heights of the integral waves are proportional to the peak area, i.e., the amount of substance present in the spot. In view of Beer's law, the proportionality factor between the absorbance and amount of substance is characteristic of the substance itself (specific absorbance), accordingly, a calibration curve is to be constructed (height of the integral wave vs. amount of the component present in the spot).

To facilitate evaluation, it is advisable to record the curves only in the vicinity of the spots and not along the whole chromatogram.

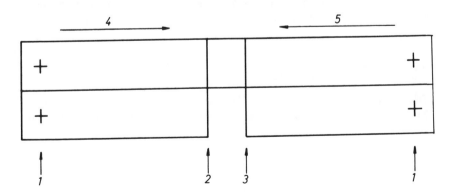

FIGURE 18.2

Preparation of the sheet before developing. 1: Start points, 2: front B, 3: front A, 4: developing B, 5: developing A.

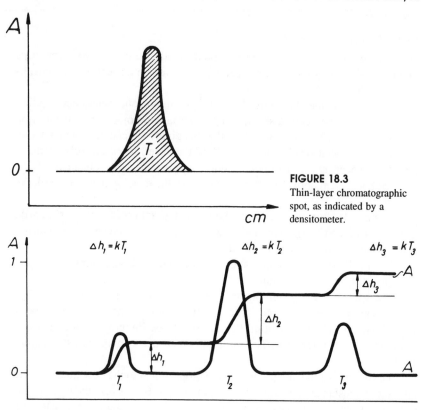

FIGURE 18.3
Thin-layer chromatographic spot, as indicated by a densitometer.

FIGURE 18.4
Thin-layer chromatographic spots as recorded by a densitometer.

Evaluation

Plot the calibration graph and read the amount of substance present in 2-, 4-, 8-, and 10-μl aliquots of the unknown. Calculate the concentration in milligrams per cubic centimeters from all the four data and take the average.

18.2. References

Fried, B. and Sherma, J., *Thin-Layer Chromatography: Techniques and Applications.* 2nd ed., (Chromatographic Science Series, Vol. 35), Marcel Dekker, New York, 1986.

Chapter 19

Gas Chromatography

Principle of the Technique _____

Gas chromatography is a method suitable for the separation, identification, and quantitative determination of the components of gas mixtures or substances which can be volatilized without decomposition. Commercially available devices can be applied to the solution of diverse problems. The nature of the column to be applied is dependent on the nature of substances to be separated. The detectors used are mostly based on ionization (e.g., hydrogen flame ionization, argon ionization, electron capture) and on heat conductivity. The electrical signal provided by the detector and recorded as a function of time is called the chromatogram, which is a recording of the result of separation. Each peak in the series of peaks represents a substance. The time elapsed from the introduction of the sample until the appearance of the maximum (retention time) or the volume of carrier gas passed through the system (retention volume) during this time is characteristic of the nature of the substance, whereas the peak area (the integral of the curve with respect to time) is in correlation with the amount of substance.

The retention volume (V_R) is the product of the retention time (t_R) and the volumetric flow rate (F) of the carrier gas (eluent)

$$V_R = t_R \cdot F \qquad (1)$$

The retention volume does not characterize the gas holdup of the column unambiguously, since it includes the eluent volume which is necessary to flush the "dead volume" between the spot of sample introduction and column, and column and detector. This "dead volume" is

$$V_M = t_M F \qquad (2)$$

223

By subtracting the dead volume from the retention volume, the reduced retention volume is obtained:

$$V'_R = V_R - V_M \tag{3}$$

The corresponding retention time is the reduced retention time t'_R:

$$t'_R = t_R - t_M \tag{4}$$

The dead time t_M is the time of the appearance of the peak of air if a heat conductivity detector is used and that of the methane peak if an ionization detector is used.

The flow rate F at the outlet of the apparatus can be corrected by taking the pressure drop along the column into consideration by using the average flow rate characteristic of the system in doing the calculations. This correction can be given according to James-Martin as the factor

$$j = \frac{3}{2} \frac{\left(p_{in}/p_{out}\right)^2 - 1}{\left(p_{in}/p_{out}\right)^3 - 1} \tag{5}$$

where p_{in} is the input pressure of the carrier gas and p_{out} is the output pressure of the carrier gas.

The corrected retention volume (V_R°) can be defined as

$$V_R^\circ = j\,V_R \tag{6}$$

If both the correction and the dead volume are taken into consideration, the net retention volume (V_N) is obtained:

$$V_N = j\,(V_R - V_M) = j\,V'_R = j\,F\,(t_R - t_{RM}) \tag{7}$$

(In the literature, usually the corrected dead volume is referred to as dead volume.)

The distribution coefficient characteristic of the substance partitioned between the flowing gas phase and the stationary phase and depending also on the temperature is defined as

$$K = \frac{V'_R}{V_S} \tag{8}$$

where V_S is the volume of the partitioning liquid (or adsorbent) in the column.

By combining Equations 3 and 8, the basic equation of gas chromatography is obtained:

$$V_R = V_M + K\, V_S \tag{9}$$

The retention data characteristic of the nature of the component can be referred to those of other substances (relative retention) or to those of several other substances, e.g., the Kováts retention indices are referred to n-alkanes.

The relative retention $(r_{1,2})$

$$r_{1,2} = \frac{V'_{R1}}{V'_{R2}} = \ldots = \frac{t'_{R1}}{t'_{R2}} \tag{10}$$

is the retention of substance 1 referred to substance 2.

The Kováts retention index (I) gives the retention data of the components referred to the retention of n-alkane homolog, taking the index values of n-alkanes to be 100 n (where n is the carbon number of the n-alkane).

The retention index of a substance is

$$I_x = 100\, \frac{\log t'_{Rx} - \log t'_{Rn}}{\log t'_{Rn+1} - \log t'_{Rn}} + 100n \tag{11}$$

The condition is that

$$t'_{Rn} < t'_{Rx} < t'_{Rn+1}$$

where t'_{Rn} and t'_{Rn+1} are the reduced retention time of n-alkanes with carbon numbers n and n + 1, respectively, between the peaks of which the peak of the component in question appears.

The index values of a great number of organic compounds measured using various columns are collected in ASTM publications edited by McReynolds. The collection can be used for the identification of peaks of unknown substances.

The efficient separation is important both in identification and determination of the components. The efficiency of the separation by a column can be characterized in terms of the resolution (R):

$$R = 2\, \frac{t_{R1} - t_{R2}}{W_1 + W_2} \tag{12}$$

and by the number of theoretical plates (N):

$$N = 16 \left(\frac{t_R}{W} \right)^2 = \frac{L}{HEPT} \tag{13}$$

where t_{R1} and t_{R2} are the retention volumes for two subsequent peaks, W_1 and W_2 are the base widths of the two peaks, L is the length of the column, and HETP is the height equivalent to one theoretical plate.

Another important parameter is the retention (or capacity ratio) which is according to a definition:

$$k = \frac{t_R - t_M}{t_R} \tag{14}$$

The value of this parameter has to be between 4 and 10 for a good capillary column.

The range of the function "integral of the detector signal vs. amount of substance" is used for the purposes of quantitative analysis where a linear relationship exists between the integrated signal (peak area) and the amount of substance:

$$A = a \cdot m \tag{15}$$

where A is the peak area, m is the amount of substance producing the peak, and a is the so-called practical sensitivity of the apparatus.

For a given detector the practical sensitivity is dependent on the nature of the substance and on the experimental conditions. The amount m_i of the unknown component can be calculated from A_i, its peak area if the practical sensitivity of the instrument for this component a_i is known. Accordingly, quantitative determinations have to be begun by determining this practical sensitivity. Various methods of quantitative analysis are distinguished according to the way of the determination of a_i.

1. Calibration method. In this method, a_i is determined directly.
2. Addition method. The determination of a_i is avoided by adding a known amount of the standard of the component to be determined to the analyzed sample.
3. Internal standard method. A relative sensitivity (f_i) referred to an internal standard is determined, using a suitable compound as internal standard which is not originally present in the sample:

$$f_j = \frac{a_i}{a_s} = \frac{A_i \cdot m_s}{A_s \cdot m_i} = \frac{M_s \, (RMR)_i}{M_i} \tag{16}$$

where the index s denotes the standard and i the component to be determined. M_s and M_i are the molecular weights, f_i is termed relative response factor in the literature, its molar value being $(RMR)_i$, the "relative molar response".

The internal standard method can be used to advantage particularly in the analysis of solutions and liquid samples, as the error involved in the measurement of volumes is eliminated. If the relative sensitivity data of the components of a mixture, referred to as an internal standard, are known, the concentration (wt %) of the i-th component can be given as

$$c_{i,x} = \frac{g_s \cdot A_i \cdot 100}{G \cdot A_x \cdot f_i} \tag{17}$$

where G is the amount (weight or volume) of the sample, g_s is the amount (weight or volume) of internal standard added, and A_i and A_s are the peak areas of the component and standard, respectively.

The sample to be analyzed is diluted with a solvent, the peak of which appears at the beginning or end of the chromatogram in order to bring the concentrations of the components within the linear range of the detector.

A special method of the internal standardization is the area normalization. In using this method, the practical sensitivity is taken into account by using an area correction factor to correct for the deviation of the area percent from the corresponding weight percent.

19.1. Investigation of the Performance Characteristics of a Fused Silica Capillary Column

Apparatus

Gas chromatograph with flame ionization detector and capillary inlet system

Column, 25 m in length, 0.33 mm in internal diameter, polymethyl siloxane stationary phase (HP-1, SPB-1, or other corresponding capillary), film thickness 0.25 μm

Microsyringes, 1 and 10 μl gas syringe

Chemicals

CH_4 for t_M measurement (from gas network)

Nonpolar text mixture (or Grob-test mixture)

Introduce 50 to 100 μl of methane into the gas chromatograph. (Column temperature: 110°C, carrier gas flow rate 1 cm^3/min, injector temperature 200°C, split ratio 1:50.) Measure the t_M value, and calculate the average linear gas velocity:

$$\bar{u} = \frac{L}{t_M}$$

It has to be equal to 20 cm/s using N_2, 30 cm/s using He, and 40 cm/s using H_2 as a carrier gas.

Then introduce 1 μl text mixture. The nonpolar test consists of 2-octanon, n-decane, 1-octanol, n-undecane, 2,6-dimethyl phenol, 2,6-dimethyl aniline, n-dodecane, naphthaline, and n-tridecane, dissolved in dichloromethane.

Evaluate the peak shape of the chromatogram according to the description of the function of the different components.

Calculate from the chromatogram the number of theoretical plate for n-tridecane and the HETP. Calculate the k values for each peak, then calculate the coating efficiency of capillary:

$$\eta = \frac{HETP_{min}}{HETP_{meas}}$$

The $HETP_{min} = r\sqrt{\dfrac{1 + 6k + 11k^2}{3(1 + k)^2}}$, where r is the radius of capillary tube in millimeters.

Calculate the separation number.

$$TZ = \frac{t_{R2} - t_{R1}}{\varpi_1 + \varpi_2} - 1$$

where t_{R2} and t_{R1} are the retention times for n-tridecane and n-dodecane and ϖ_1 and ϖ_2 are the peak widths at the half height.

19.2.   Identification of Alkyl-Benzene Homologs

Apparatus

> Gas chromatograph with flame ionization detector and capillary sample inlet
> Column, 25 m × 0.33 mm, polymethyl siloxane with 0.25-μm film thickness
> Microsyringes

Chemicals

> Unknown sample containing alkyl-benzene homologs
> CH_4
> Mixture of n-alkanes (n-C_7-n-C_{12}) dissolved in n-hexane

Introduce 50 to 100 μl of methane into the gas chromatograph. Measure the t_M value. Then introduce 1 μl of n-alkane's mixture and unknown sample separately. (The injector temperature 200°C, the column temperature 100°C, split ratio 1:50, and the carrier gas flow rate has to be the optimal value.) Measure (or read from the hard copy of the integrator) the retention data. For the easy calculation make a table:

t_R	$t'_R = t_R - t_M$	$\log t_R$	I
n-C_7			700
n-C_8			800
.			.
.			.
Unknown$_1$			I_1
.			.

Calculate the Kováts-retention indices using Equation 11. Look for the unknown substances in an index compilation or database system (e.g., ASTM, Sadtler) by means of calculated indices.

19.3. Quantitative Analysis of a Hydrocarbon Mixture

Apparatus

Gas chromatograph with flame ionization detector
Microsyringes, 1 and 10 μl
Column, 2 m in length containing SE-30 or Silicone Oil 550 as partition liquid or apolar fused silica capillary column, 25 m × 0.33 mm, with 0.25-μm film thickness
Vessels for the samples, closed by means of silicone rubber caps
Glass tubes with drawn-out end for adding chemicals

Chemicals

n-Hexane (solvent)
n-Octane
n-Nonane (internal standard)
n-Decane
n-Undecane
n-Dodecane, etc. as standards
Hydrocarbon mixture of unknown composition

Procedure

Prepare a mixture of the standards similar in composition to the sample to be analyzed, by weighing. Calculate the composition of the mixture prepared in percent by weight (c_i wt %).

Introduce 0.1 to 0.5 μl of the mixture into the gas chromatograph. (Temperature of the column: 180°C, flow rate of carrier gas, nitrogen: 50 cm³/min.) Measure the peak areas on the chromatogram and calculate the peak area percent values (c_i a %). The peak area correction factors of the components are calculated from the weight percent data:

$$t_i = \frac{c_i \ \text{wt} \ \%}{c_i \ \text{area} \ \%}$$

Run the chromatogram of the unknown, and calculate the concentration (wt %) of the components by taking the peak area correction factor into consideration

$$c_{i,x} = \frac{t_i \cdot A_i}{\sum\limits_{j=1}^{n} A_j} \ 100$$

Then, determine the composition of the unknown using n-nonane as internal standard. First, determine the f values for the components, referred to n-nonane. Weigh 0.1- to 0.3-g portions of the hydrocarbon standards (m, m...etc.) and of n-nonane (m) into a vessel, using an analytical balance, and add n-hexane in such an amount that the concentration of the components does not exceed 1 to 2% by weight.

Calculate the f values for the components from the peak area values and weights, using Equation 16.

Weigh about 0.5 g (G) of the unknown and 0.1 g of the internal standard (g) into a vessel, and add n-hexane. Introduce 0.1 to 0.5 μl of this solution into the gas chromatograph.

Calculate the composition of the mixture from the peak areas using Equation 17.

Evaluation
The composition of the unknown (wt %) is given, as determined

1. By the normalization method
2. By the internal standard method
3. Calculate the deviation of the results obtained by the two methods, in percent

19.4. Determination of Alcohols in Plum Brandy

Apparatus

Gas chromatograph with accessories
Sample containers equipped with silicone rubber closure caps

Hamilton syringe, 10.0 μl
Ruler, subdivided 0.5 mm
Column, 25 m × 0.33 mm, stationary phase immobilized
Carbowax 20 M, film thickness 0.25 mm

Chemicals

Methyl alcohol
Ethyl alcohol
i-Propyl alcohol
n-Propyl alcohol
i-Butyl alcohol
n-Butyl alcohol
i-Pentyl alcohol
n-Pentyl alcohol (internal standard)

In the first step, the relative sensitivity of alcohols, referred to as the internal standard, is determined. For this purpose 0.1 g from each alcohol is weighed to an accuracy of four decimals into a 10-cm^3 volumetric flask. The flask is filled up to 10 cm^3 with distilled water. Of this mixture 0.5 or 1 μl is injected into the gas chromatograph (under the following conditions: injector temperature 200°C, heated FID temperature 200°C, column temperature is programmed: initial temperature 80°C during 4 min, then 12°C/min until 140°C, then isotherm on this final temperature during 5 min). Calculate the relative sensitivities using Equation 16.

In the second part of the exercise measure 0.5 to 1 g of plum brandy and 0.1 g n-pentyl alcohol by weight into a volumetric flask of 25 cm^3 and filled up to 25 with distilled water. Introduce 1 to 2 μl into the apparatus from this mixture. On the basis of the peak areas calculate the concentration of alcohols using Equation 17.

19.5. References

Hyver, K. J., Ed., *High Resolution Gas Chromatography,* 3rd ed., Hewlett-Packard Co., 1981.

Perry, J. A., *Introduction to Analytical Gas Chromatography: History, Principles and Practice,* (Chromatographic Science Series, Vol. 14), Marcel Dekker, New York, 1981.

Chapter  20

Liquid Chromatography

20.1. Evaluation of Column Efficiency and Selectivity

Theory

The stationary phases used in high-performance liquid chromatography (HPLC) are prepared in a different way. The column efficiency and selectivity depends on several factors. For example, in reversed-phase HPLC the efficiency and selectivity depend not only on the chain length of alkyl modifier but several other factors, i.e., base silica gel structure. The evaluation of a reversed phase or other stationary phase depends on a practical purpose, namely, what application the method is used for. The first step before using a reversed phase or other column is to check the quality the packing.

In the following discussion we summarize the general terms used in column evaluation. Checking the quality of column is based on the Knox[1] equation:

$$h = Av^{1/3} + \frac{B}{v} + Cv$$

where   h = reduced plate height

n = reduced velocity

A, B, C = constants

In this case we control only kinetic efficiency of a reversed-phase column. For a good column we get the following numbers A = 1 to 2, B = 1, and C = 0.04.

In the first part of this experiment the kinetic parameters will be evaluated. The expressions used in this experiment are

$$N = 16\left[\frac{t_R}{w}\right]^2$$

where N = plate number
       t_R = retention time
       w = peak width at baseline

or

$$N = 5.54\left[\frac{t_R}{w_{1/2}}\right]^2$$

where $w_{1/2}$ = peak width at half height

These equations are valid for symmetrical Gaussian peaks. This assumption is generally valid for test solutes used for controlling the packing quality of columns. For a non-Gaussian peak the method of Foley and Dorsey[2] is used. In this case, the equation used is

$$N = \frac{41.7\left[\frac{t_R}{w_{0.1}}\right]^2}{1.25 + \dfrac{b}{a}}$$

where $w_{0.1}$ = peak width at 10% of peak height
       $\dfrac{a}{b}$ = asymmetry factor

The quantities used in the Dorsey-Foley[2] method are shown in Figure 20.1.

The elimination of the dependence of N on column length can be calculated by the plate height:

$$H = \frac{L}{N}$$

where L = column length

The Knox-equation is derived from the well-known van Deemter equation. The simplified form of this equation is

$$H = A + \frac{B}{u} + (C_s + C_m)u$$

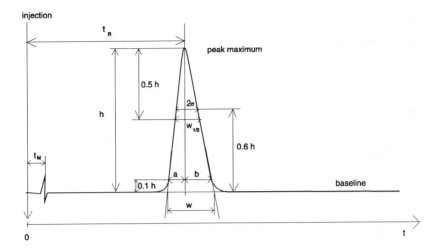

FIGURE 20.1
Determination of the quantities used in the equations for calculation of efficiency of an HPLC column.

where A = convective dispersion factor
B = axial dispersion factor
C_S = mass transfer resistance in the stationary phase
C_m = mass transfer resistance in the mobile phase
u = linear velocity of the mobile phase

The linear velocity of the mobile phase can be calculated from the column length and the retention time of an unretained solute:

$$u = \frac{L}{t_M}$$

where t_M = retention time of an unretained solute

The meaning of t_M in liquid chromatography is not clear. There are several methods for its determination, e.g., in RP-HPLC injecting an aliquot of the mobile phase with a slightly different composition or a sample contains $NaNO_3$ or uracil, etc. The choice of the unretained solute depends on the conditions used in RP-HPLC.

The use of the Knox equation for evaluation of chromatographic columns allows the comparison of columns packed with different particle sizes and operated with different linear velocities. The calculation of variables can be done using the following expression:

$$h = \frac{H}{d_P}$$

where $\bar{d}_P$ = average particle diameter

and the reduced velocity (ν)

$$\nu = \frac{ud_P}{D_M}$$

where D_M = diffusion coefficient of the measured solute in the mobile phase

When an accurate value for D_M is not available an approximate one can be calculated by the Wilke-Chang equation:

$$D_M = \frac{A(\Psi M_2)^{\frac{1}{2}} T}{\eta V}$$

where A = constant, (typically 7.4×10^{12}, if viscosity is expressed in m^2/s)
        M_2 = molecular weight of solute
        Ψ = solvent dependent constant for methanol 1,4 and 2,6 for water
        T = temperature in K
        η = the solvent viscosity
        V = molar volume of solute

For mixed solvents (biner eluents) the nominator for low molecular weight solutes fall in the range 0.5 to 3.5×10^{-9} m^2/sec.[3,4]
    The column flow resistance parameter gives information about the resistance of flow of a given column and mobile phase used:

$$\Phi = \frac{\Delta p \, \bar{d}_P \, t_M}{\eta L^2}$$

where Δp = pressure drop on the column used

From the column resistance parameter a new parameter can be calculated:

$$K = \frac{\bar{d}_P^2}{\Phi}$$

The separation impedance E, introduced by Bristow and Knox[1]:

$$E = h^2 \Phi$$

The less the separation impedance, the better the column used in high-performance chromatography. It means low reduced plate height and low column flow resistance. For a good conventional packed column the value of reduced plate height is about 2, the flow resistance parameter is 500, and the separation impedance is 2000.

20.1.1.   Evaluation of Column Efficiency and Selectivity in Normal Phase Chromatography

Some General Remarks

A new column must be tested before use and periodically retested during its use. Good column performance parameters are the first indication for column selection but one must not forget that the selected test compounds differ from the solutes. There are several indications which show that different stationary phases have different character; even in a test procedure they give the same performance. It is well known that for measuring of primary kinetic parameters some desirable properties of test solutes and mobile phases must be fulfilled:

* Low molecular weight and rapid diffusion between the two phases (mass resistance, kinetic resistance as low as possible).
* Mobile phase has to be low viscosity and simple composition.
* Solutes have to represent different molecular interactions to characterize the column in terms of thermodynamic efficiency to some extent.
* Injected volume and amount must be lower than overload value of the column.
* The capacity ratio of solutes should be between 0.1 and 10.
* Molar absorptivity of test solutes should be adequate for measurement at 254 nm.

For the determination of thermodynamic performance one must know the column dead volume. In all cases the method of determination of dead volume must be given. In a practical point of view, the sample which has slightly different composition of mobile phase can be injected into the column, and the detector wavelength must be adjusted as low as possible; the first peak retention time (negative or positive) can be considered as t_M and column dead (void, holdup) volume can be calculated; the test mixture contains a solute which is practically unretained.

To get comparable results the same set of solutes and mobile phase composition must be used for checking the column performance time to time.

For determination of column kinetic and thermodynamic performance we suggest the use of a mixture containing toluene/nitrobenzene/2,6-dinitrobenzene/1,3,5-trinitrobenzene, and eluent composition is hexane/methanol 99.5/0.5 v/v.

Apparatus

Isocratic chromatographic system equipped with UV detector and column filled with silica gel
Graduated cylinders, 500 cm^3 and 10 cm^3
Graduated flasks, 25 cm^3

Solvents and Chemicals

Hexane, methanol, methylene chloride, 2-propanol, chromatographic grade
Toluene, nitrobenzene, acetophenone, 2,4-dinitrotoluene, 1,3,5-
trinitrotoluene

Procedure

Before testing a new or used column, one has to clean and activate it. For this
purpose, prepare an eluent consisting of hexane/methylenechloride/2-propanol
40/50/10 v/v and wash the column, using about 100 cm^3, at a flow rate of 2 to 3
cm^3/min. For activation use pure methylene chloride, about 100 cm^3, at a flow rate
of 3 to 5 cm^3/min. After cleaning and activating the column, prepare the eluent for
the testing procedure. Take 497.5 cm^3 hexane and 2.5 cm^3 methanol separately and
pour into the eluent container and mix them. For a few minutes, purge the pump
with eluent and set the flow rate to 2 cm^3/min. The equilibrium can be reached in
about 30 to 40 min. To be sure, inject the test mixture twice. If the two retention
times of test solutes are within 2%, the experiment can be started.

Set the flow rate at 2.5, 2, 1.5, 1, 0.5, and 0.2 cm^3/min and inject the
text mixture at least twice. The text mixture concentration is for toluene
0.2, nitrobenzene 0.010, acetophenone 0.025, 2,4-dinitrotoluene 0.010, and
1,3,5,-trinitrotoluene 0.010 mg/cm^3. Injection volume is 20 µl.

Evaluation

From a Simple Chromatographic Run

In this case we can get fast information on efficiency and selectivity and about
the column condition after use.

The parameters calculated from the chromatogram are net retention times
$(t_N) = t_R - t_M$ and volumes, capacity ratios, plate number of the less retained and
the last chromatographic peak, asymmetry factor compounds mentioned above,
and the separation factors.

If no instrumental contribution is present for the asymmetry factor, the
value is less than 1.2 for $k < 1$ and 1.8 if $k > 2$ is acceptable. If the value is out
of this range the column is poorly packed. If only the unretained peak is
asymmetric and there is a great difference between the plate number of unretained
and retained component, the extra column effect might cause this (improper
connections, etc.).

The plate number depends on the average particle diameter. For unretained
or weakly retained compounds, the absolute value for H is about twice the
particle diameter; for retained solutes ($k > 2$) the absolute value is two to five
times greater than $\bar{d}_p$.

For a column packed with 5 µm silica gel the H is about 10 µm for
unretained solute. The N/[m] is 100,000; for a retained compound it is only
about 50,000. These values are valid only for test compounds without any

strong, specific interactions with the stationary phase. If the plate number falls below this theoretically estimated value, it means a poorly packed column. If the plate number drops during the run and peak asymmetry increases greatly, the column should be discarded. If the capacity factor and separation factor (selectivity) values for the test solutes change during the chromatographic run, this indicates a change in the nature of the packing material. In normal phase chromatography all solvents must be free from water or water content must be controlled very strictly; otherwise, changes in capacity factor and separation factor are due to the different water content. In such cases, we have to activate the column and the test procedure must be repeated.

If the pressure drop will change during the run, the column resistance factor will increase. One reason for this increase is that the fittings or inlet or outlet filter are partially blocked. If after cleaning the fittings and replacing the inlet and outlet filter the pressure drop does not restore to the normal range, the column should be discarded.

Complex Evaluation of Column Performance

From a single chromatographic run we can conclude the column main parameters and we can decide the use of the column. To get more detailed information of chromatographic performance we have to calculate the following parameters and relate to running conditions and column parameters. Parameters must be calculated: N, N_{sys}, H, h, v, v, Φ, E, K.

The plot of the plate height against the linear velocity and the reduced plate height against reduced velocity must be constructed in log-log form. Some general remarks have been taken based on values N and H in "evaluation from simple chromatographic run"; they are valid for all solutes taken into consideration.

The reduced plate height for a conventional packed column of about 2 is considered excellent. In practice a value of 3 is acceptable. The flow resistance parameter will normally be between 500 to 1000. If this value is ten times higher, a partial blockage in the chromatographic system can be suspected.

In the Knox equation the constant A is a measure of how well the column was packed. The optimum value for A is about 1; for a poorly packed column the value is 2. There is no general agreement about constant C value. Different authors have published different values. Unger[5] stated an optimum value is about 0.1; according to Poole[6] this value for porous packing is 0.05.

20.1.2. Evaluation of Column Efficiency and Selectivity in Reversed-Phase HPLC

Column Activity Tests for Chemically Bonded Phase

Theory

Chemically bonded phases are produced by reacting the accessible silanol groups on the surface of silica gel with a reactive organochlorosilane reagent.

The most frequently used chemically bonded phases are octyl or octadecyl functions covalently bonded to the silica surface. The process used to prepare different silica gels, functionality of alkyl-silane used, and conditions in preparation influence the performance of the chemically modified silica gel. It is well known that only a part of silanols can be reacted by modifiers. The maximum conversion will be determined by the cross-sectional area of alkyl-chlorosilane. Not more than half of silanol groups can be reacted. Some parts of unreacted silanols can interact with solutes investigated and cause great differences in selectivity between stationary phases. Residual trace metal impurities can influence the kinetic and thermodynamic performance of chemically bonded phase. Sodium, aluminum, and iron have been determined by atomic spectroscopy.[7] At least three factors influence the kinetic and thermodynamic performance of chemically bonded phases: hydrophobicity of stationary phase, accessible silanol groups, and trace metal impurities. Not mentioned here are the pore structure of bonded phase, polymeric character of bonded layer, etc.

The hydrophobicity of a column can be measured by retention of nonpolar solute under standard conditions. Some proposed test mixtures are anthracene/benzene eluent: acetonitrile/water 64/36 v/v,[8] pentylbenzene/butylbenzene eluent: methanol/water 4:1 v/v,[9] and ethylbenzene/toluene eluent: methanol/water 65/35 v/v.[10]

Walters[8] characterized different octadecyl modified stationary phases and large differences in the hydrophobicity index were found. There is only a limited number of data about lot-to-lot variation according to hydrophobic character of stationary phases produced by a single manufacturer.[11] A lot-to-lot variation depends on the choice of compounds used in the procedure of column evaluation.

The nature of the bonded phase depends on the reagent (mono-, di-, or trifunctional chlorosilane used) and the conditions. A simple test has been developed to determine the bonding chemistry used to prepare the octadecylsiloxane stationary phases. The retention behavior of polycyclic aromatic hydrocarbons (PAH) depends on structure of the bonded ligand. It was found that, if the column is monomer (brush-type), the retention order benzo(a)pyrene (BaP) < phenanthrophenanthrene (PhPh) < 1,2:3,4:5,6:7,8-tetrabenzonaphthalene (TBN), eluent was acetonitrile/water 85/15 v/v, on polymeric phase PhPh < TBN < BaP. For densely bonded monomer or lightly loaded polymeric phases the elution order of test solutes is PhPh < BaP < TBN. Based on the PAH retention, the columns can be classified into three groups.[12-16] If the separation factor for TBN and BaP is between 1.7 and 2.2 the stationary phase is a monomer type; if $1.0 < \alpha < 1.7$ the stationary phase is intermediate; if $0.5 < \alpha < 0.9$ the stationary phase is a polymeric type. This procedure is called column shape selectivity assessment.

Apparatus and Chemicals

The apparatus and chemicals are the same as those used in the basic evaluation of the stationary phase. The eluent is 85/15 v/v acetonitrile/water; the test mixture concentration is about 1 to 10 µg/cm^3. The solvent composition for test mixture is about the same as that for eluent composition. Dissolve the PAH (TBN and BaP) in different volumetric flasks; first add acetonitrile into the volumetric flasks and put into an ultrasonic bath for a few minutes and adjust the volume with distilled water.

Procedure

Prepare the eluent by measuring 425 cm^3 chromatographic grade acetonitrile and 75 cm^3 distilled water, separately. Degas the eluent by placing it in an ultrasonic bath or purge by helium. Purge the pump about 5 min with eluent. Adjust the flow rate to 1 cm^3/min and let the instrument run for about half an hour. Adjust the wavelength to 254 nm and inject 20 µl onto the chromatographic column from PAH solutions. From every test mixture inject at least twice.

Evaluation

According to empirical test developed by Sander and Wise if a separation factor for TBN and BaP lies between 1.7 and 2.2 the stationary phase is monomer; if $1.0 < \alpha < 1.7$ an intermediate; if $0.5 < \alpha < 0.9$ a polymeric stationary phase is used.

Test for Column Determination During Use

The instrumentation and chemicals are the same as those used before. The test mixture is anthracene/benzene; the concentration is 2 to 10 µg/cm^3 for anthracene and 10 to 50 µg/cm^3 for benzene. The final solvent composition for the text mixture is about same as the eluent one. To be sure the anthracene will dissolve, first put acetonitrile to solute in an ultrasonic bath (a few minutes is enough) and adjust the volume by distillate water. Eluent composition is 65/35 v/v acetonitrile/water.

Procedure

Prepare the eluent by measuring 325 cm^3 acetonitrile and 175 cm^3 distillate water separately and put into the eluent container and degas it (ultrasonication, vacuum, or helium purge). Purge the pump about 5 min, and adjust the flow rate to 1 cm^3/min; let the instrument run about 30 min. Adjust the detection wavelength to 254 nm, and inject 20 µl sample and repeat it at least twice. If the retention times of test solutes differ by only 1 or 2%, the average values are used for evaluation. If the deviation in retention times is higher, repeat the injection until the difference is lower.

Evaluation

From time to time repeat the measurement and record the data. Calculate the separation factor for anthracene and benzene. If the retention data and separation factor decrease significantly, the carbon load of stationary phase will decrease simultaneously and the column may be discarded.

Determination of Residual Silanol Groups on the Surface of Chemically Bonded Phases

Theory

The residual and accessible silanol groups cause peak tailing, irreproducible retention times. These undesirable effects are particularly prevalent for amines and strong hydrophobic bases. Several methods are used by the manufacturer to reduce this effect, but none of them can eliminate the interaction totally. For example, special columns are offered to determine basic drugs. Unfortunately, during the chromatographic run the octadecyl or other bonded ligand can be removed from the surface of silica gel. It generates an easily accessible silanol group, which changes the retention and selectivity of column. The residual silanol groups are not homogeneous with respect to the molecular interaction and its energy.[11,17,18]

A small portion of these are highly acidic and mainly responsible for the peak broadening of ionized solutes under the condition of chromatographic run. On these silanol groups the proteins can be adsorbed irreversibly and probably denatured. The metal impurities on the surface of the chemically bonded phase might activate the silanol groups. Both the activated silanol groups and the metal impurities can strongly interact with solute ability to form chelate; the result is peak broadening, tailing, and nonreproducible retention time.[19,20] To reduce the influence of metal impurities on retention and peak broadening, some manufacturers started to supply metal-free stationary phases. Several methods are suggested to characterize the residual silanol groups on the surface of chemically bonded phases.[11,21-27]

One main group of test systems employs normal phase conditions;[21,22] another one uses water-miscible solvents as the mobile phase.[23-27] If normal phase conditions are used the time to remove the water from the column is relatively high. This is an easier and faster test procedure when reversed-phase conditions are used.

A silanophilic index[26] was defined by Sadek:

$$R = \frac{k_{cyclam} - k_{chrysene}}{k_{chrysene}}$$

Others define the silanophilic index based on the separation factor of a polar and an apolar test solute.[11,27] Kimata used the separation factor of benzylamine/phenol test solutes to measure the ion exchange capacity of the reversed-phase stationary phase.[27] It was found that the separation factor varies from 0.01 to 1.43, indicating a wide range of highly active silanol groups is on the surface of a modified silica gel.

Apparatus

Isocratic chromatographic system (pump, injector, UV detector, data handling device)
Volumetric flask, 25 cm^3
Graduated cylinder, 500 cm^3
Ultrasonic bath

Chemicals

Distilled water, methanol
Aniline, 10 μg/cm^3
Phenol, 10 μg/cm^3

Procedure

Preparation of eluent: pour 167 cm^3 chromatographic grade methanol into a graduated cylinder and pour 333 cm^3 distilled water into another one. Mix them in an eluent container and place into the ultrasonic bath for a few minutes for degassing (it can be used for other degassing methods as well). Purge the pump about 5 min and adjust the flow rate to 1 cm^3/min. Run the instrument about 30 min, and adjust the detection wavelength to 220 nm. Inject 20 μl into the chromatographic column at least twice. For determination of t_M inject the solution of 20/80 v/v methanol/water and measure the disturbance at its lowest wavelength (about 205 nm or lower).

Evaluation

From retention time test solutes measured and phase disturbance, calculate the separation factor. If the separation value is low the column is good with respect to the accessible highly active silanol groups. If the value increases during usage of the column phase, deterioration is probable.

Checking of Metal Impurities

Theory

It has been described that metal impurities can activate the silanol groups and strongly interact with solutes to form chelate complexes. The 2,4-pentanedione can form a strong complex with different cations. For a metal-free packing no retention of 2,4-pentanedione can be observed. The higher the metal content, the greater the retention time of 2,4-pentanedione and gradually the deterioration of the peak shape. If no peak is observed for 2,4-pentanedione, the surface of the stationary phase is highly loaded by metal impurities. Therefore, the retention characteristic of 2,4-pentanedione is a sensitive way to estimate the trace metal content of the column.[24,28]

Apparatus

Isocratic HPLC system with UV detector and data handling
system
Ultrasonic bath
Graduated cylinder, 500 cm^3
Filtering device, with 0.5 μm membrane filter

Chemicals

Methanol chromatographic grade
Double distillate or highly purified water
Sodium acetate
2,4-Pentanedione (10 μl/cm^3 in eluent)

Procedure

Preparation of eluent: 300 cm^3 methanol and 200 cm^3 water are measured
into a graduated cylinder, separately. Into the water put 2.5 g anhydrous
sodium acetate; dissolve and filter with about 0.5-μm pore diameter membrane filter. Mix water and methanol in the eluent container and degas it,
placing it into an ultrasonic bath for a few minutes (degassing can be done
by vacuum or helium purge). Purge the pump about 5 min, adjust the
wavelength to 254 nm, run the instrument about 30 min, and inject the test
solution at least twice. Determine the holdup volume (void volume, dead
volume) by injection 50/50 v/v methanol/water solution at the lowest wavelength as possible (phase disturbing).

Evaluation

Calculate the capacity factor of 2,4-pentanedione and plate number. If there is
a k value of about 0, metal impurities on the surface of bonded phase are
negligible.

General Efficiency Test Method for Testing of Column Performance

Apparatus

Isocratic liquid chromatograph consisting of
a pump pulsation <1%;
an injection valve equipped with 20-μl sample loop;
UV detector, usually a variable wavelength UV or UV-VIS
detector;
a recorder or integrator or any data handling system

Ultrasonic bath
Pipettes 5 cm^3
Volumetric flask 25 cm^3
Micro-syringe for injection, 50 μl

Solvents and Chemicals

Acetonitrile chromatographic grade
Distillated or high purity water
Resorcinol, acetophenone, naphthalene, anthracene test mixture

Procedure

The liquid chromatograph is set in operation according to the Instruction Manual with eluent acetonitrile/water 55/45 v/v, flow rate 1 cm^3/min. The eluent preparation must be done the following way. Take 275 cm^3 acetonitrile and 225 cm^3 high purity water, separately, put it into the eluent container, and remove the dissolved oxygen by placing it into an ultrasonic bath for a few minutes. Before starting the experiment the chromatographic system must be run for at least half an hour.

During the equilibration of liquid chromatographic system, prepare the test mixture. Weigh 15 mg resincol, 6 mg acetophenone, 5 mg naphthalene, and 3 mg anthracene and put into a volumetric flask, nominal volume 25 cm^3. Pour about 20 cm^3 acetonitrile into the volumetric flask, shake or place into an ultrasonic bath until all components will be dissolved, and adjust the volume to 25 cm^3 with distillated water; mix the solvents.

Adjust the detector wavelength to 254 nm and sensitivity of about 0.1. Take a 50-μl sample from the test solution and inject into the chromatographic column, repeat it at least twice.

Adjust the flow rate to 0.2, 0.5, 1.5, and 2 cm^3/min and inject the sample twice at each flow rate.

For the determination of column void volume or mobile phase volume inject onto the column KNO_3 solution at 215 nm, about three times.

Evaluation of Results

1. The following parameters must be calculated for measured compounds: t_N, k, V_R, V_N, N, n, N_{sys}, H, h, d, Φ, k, E, u according to the equations given.
2. H-u curve must be drawn from data.
3. lgh-lgu curve must be drawn from data.
4. Compare the data published. These results are characteristic for column efficiency.
5. Calculate the α and R_S which are measures of the thermodynamic efficiency of the column.
6. α-F and R_S-F curves must be drawn.

20.1.3. Evaluation of Column Efficiency and Selectivity in Ion Pair Chromatography

Theory

Ion pair chromatography is a good example of the use of secondary chemical equilibrium (SCE) to control retention and selectivity in HPLC. The term secondary equilibrium is used for all other equilibria which happen simultaneously with the primary equilibrium. The primary equilibrium is the distribution of the solute between the mobile phase and stationary phase. In SCE at least two molecular forms are present in mobile and/or stationary phases. The different molecular forms have different polarities, and most of the cases have different distributions characteristic in the chromatographic system. If the equilibrium between different molecular forms is rapid enough, then only one peak will elute. If equilibrium between different molecular forms is slow, the chromatographic peak either broadens and/or multiple peaks would be observed. Controlling the different molecular forms is one way to control the efficiency and selectivity in HPLC. In aqueous and mixed organic-water solutions which are used in RP-HPLC, a large amount of data has been known for secondary equilibrium; in addition to this an easier application and rapid column equilibrium have resulted in RP-HPLC becoming dominant in SCE.

There are several forms of SCE (ion suppression, argentation chromatography, etc.). In our discussion we deal only with ion pair chromatography, which is very popular and can be used for a wide range of solutes analyzed.

An exact definition for ion pair chromatography is still lacking. It is well known that in any given liquid phase oppositely charged ions can attract each other. Depending upon the dielectric constant of the liquid, type of the ions, molecular interaction and solvation, an ion association occurs. If the two oppositely charged ions can be sufficiently bound to one another, a partly or fully neutralized species can be formed. Formation of ion association is called ion pairing.

The formation of an ion pair depends upon of dielectric constant of the medium,[29] Van der Waals interaction of oppositely changed ion,[30] dipole interaction,[31] and hydrogen-bridge interaction. In the case of RP-HPLC coulomb attraction is not sufficient to form a stable ion pair in water or organic-water mixture.[32] A large hydrophobic ion can interact with the nonpolar portions of the ions to associate via hydrophobic interaction.

According to the solvation of ions, which depends on the dielectric constant of the medium, two types of ion pairs can be formed: tight or contact ion pairs and loose or solvent-separated ion pairs.[33,34] The ion pair formation in a medium can be envisioned by a two-step process:[35]

$$A^+ + B^- + S \Leftrightarrow \left(A^+, S, B^-\right) \Leftrightarrow \left(A^+ B^-\right) + S$$
$$\text{loose} \qquad\qquad\qquad \text{tight}$$

where S is the solvent molecule

Tight or contact ion pairs can be formed by repulsion of solvents. The ion chromatographic point of view is that the tight or contact ion pairs would be favorable, because one molecular form is present only and kinetic efficiency will be high. If the solvation energy of an ion is high, solvent molecule is built in between the two ions, which are forming the ion pair. Pioneer work has been done by Shill, who adopted the ion pair extraction to modern HPLC.[36,37] The underlying idea is that in a two-phase system the ions will be predominantly found in the aqueous phase and neutral molecules can solve by organic phase; the distribution constant can be derived the following way:

$$A_{aq}^+ + B_{aq}^- \Leftrightarrow \left[A^+, B^-\right]_{org} \qquad E_{AB}$$

where E_{AB} is the extraction constant and the corresponding distribution ratio of solute A, D_A is

$$D_A = E_{AB} \left[B^-\right]_{aq}$$

where B is the concentration of ion pairing

The capacity factor in chromatography can be given by D_A:

$$k = \Phi E_{AB} \left[B^-\right]_{aq}$$

where Φ is the phase ratio (V_s/V_M).

From this simplified picture some basic feature can be concluded. In the linear part of adsorption isotherm of ion pairing reagent the capacity factor will increase with increasing concentration of ion pairing reagent.

If the logarithm of retention factor (ln k) is plotted against the logarithm of the concentration of the ion pairing reagent, the mobile phase gives a fairly straight line or slightly curved but monotone line. This approach can be used in many applications. In practice, the capacity factor reaches a plateau and may slowly decline at a still higher concentration.[38]

If we consider the solute concentration, which does not participate in the ion pair formation, then its distribution between the two phases can be expressed as follows:

$$D_A = \frac{\left[A^+\right]_s + \left[A^+, B^-\right]_s}{\left[A^+\right]_m + \left[A^+ B^-\right]_m}$$

Using this equation we can express the capacity factor:

$$k = \Phi \frac{K_0 + E_{AB}\left[B^-\right]_m}{1 + K_1\left[B^-\right]_m}$$

where K_0 is distribution constant of solute between the mobile and the stationary phase

$$E_{AB} = K_1 K_2$$

where K_1 is the ion pair formation constant in the mobile phase and K_2 is the distribution constant of ion pair to the stationary phase.

This equation can describe the plateau and slow decline of the capacity factor with increasing ion pair concentration.

E_{AB} largely depends on ion pair hydrophobicity and the hydrophobic surface interacted with solute. In a homologous series of ion pairing reagents, the longer the alkyl chain, the higher the retention of a given solute. It has been shown that each methylene group increases E_{AB} by a factor of 3 to 4[39] and, of course, the retention factor E_{AB} is influenced by the geometry of the ion pairing reagent, giving lower retention compared to a long-chain one (the number of carbon atoms are equal).

Organic modifiers have different dielectric constants and through this can effect both the retention and selectivity. Both the formation of tight or loose ion pair and adsorption of ion pairs and ion pair reagents depend upon the organic modifier itself and the concentration in the eluent. Others have stated the concentration of ion pairing in mobile phase and the organic modifier have the greatest effect on adsorbed amount on the surface of reversed-phase stationary phase.[43-45] If the surface coverage by ion pair reagent is constant, the retention and selectivity is hardly influenced by other factors.[42,46,47]

At the eluent pH the solute and ion pairing reagent must be ionized to get a high value for E_{AB}. Using silica-based reversed-phase packing the working pH range is about 2 to 8 for polymeric material there is no pH limit (pH range between 1 and 14). The pH is controlled by adding a suitable buffer to the eluent. Inorganic phosphate, citric-acid-citrate, acetic-acid-citrate, and alkyl-amine-phosphate can be used in different concentrations.[43,48] The buffers can be soluble in the eluent used; at high organic modifier concentration, solubility of the organic buffer is better. The buffer concentration using ion pairing chromatography is between 0.5 and 50 mM and contributes to the ionic strength of the eluent. Strong acid and base are ionized throughout the pH range used in RP-HPLC. Weak acids ($pK_a > 3$) are ionized above pH > 5, and their retention and selectivity depend upon the type of ion pair used; in the pH range 2 to 5 ionization of solutes is suppressed and the influence of ion pairing reagent is small; the retention and selectivity depend upon solute polarity. Weak bases ($pK_a < 8$) are fully ionized at pH below 5, and their retention is dependent upon the type and nature of ion pairing reagent used. At pH above 6 the ionization of weak bases is suppressed and retention and selectivity depend upon the polarity of weak bases.

In ion pairing chromatography the ion strength can control the retention and selectivity. The buffer ions may compete with the counterions for the charge sites of absorbed ion-pairing reagent (ion exchanger mechanism). The solubility of the solutes and ion pairing reagent is affected by ionic strength due to salting-out effects.[49,50] Adsorption of ion pairing reagent on the surface of reversed-phase stationary phases can be affected by changes in ionic strength. According to dynamic ion exchange model[51-55] the number of absorbed ion pairing reagent will determine the number of ion exchanger sites and of course the retention as well. It is recommended to keep the ionic strength constant by adding varying amounts of the neutral salt.[56]

The temperature will affect both retention and selectivity in ion pair chromatography, but we use this separation method at ambient temperatures for most of the application.

Parameters controlled in reversed-phase ion pair chromatography are

- Type and concentration of ion pairing reagent
- Nature of ion pairing reagent (change length, bulky structure, salt or ionizable)
- Organic modifier(s) its concentration in the mobile phase
- Buffer type and its concentration used for adjusting pH
- Ionic strength (salt concentration)
- Temperature

To understand the retention and selectivity several theoretical models have been proposed. They can be classified into three main groups: the ion pair model,[38,54,57,58] the dynamic ion-exchange model,[51-55] and the electrostatic model.[59-64]

The ion pair model assumes that an ion pair formed in the mobile phase and the ion pair can adsorb on the surface of reversed-phase stationary phases. The capacity factor is governed by ion pair formation constant (K_1) in the eluent and extraction of ion pair complex into the stationary phase (K_0, E_{AB}). This model can be used successfully in liquid-liquid chromatography, but it has some disadvantages when applied to chemically bonded phases.

In the dynamic exchange model similar treatment is used as in ion exchange chromatography. Retention of ionized solutes is mainly governed by ionic interactions. The ionic interactions are dependent upon a surface charge which can be calculated from an adsorbed amount of ion pairing reagent. Parameters affecting the adsorption of ion pairing reagent result in a change of retention and selectivity.

The electrostatic theory of ion pair chromatography is based on the classical electrostatic model. If the charged molecules are adsorbed onto a surface the oppositely charged ions will associate to the adsorbed layer to maintain electrical neutrality. The formation of electrical double layer results in a difference in electrostatic potential between the bulk solvent (mobile phase) and the surface of the stationary phase. If an ion pairing reagent has high affinity (long alkyl chain) to hydrophobic surface (alkyl-modified silica gel), the difference in electrostatic potential between the two phases is higher. The surface potential depends on the adsorbed amount of ion pairing reagent, dielectric constant of the eluent, and ionic strength. For the retention an equation can be derived:

$$k = \Phi \exp\left(-\Delta G^0/RT - zF\Psi_0/RT\right)$$

where Φ is chromatographic phase ratio, z is charge of the solute, Ψ_0 is the difference in electrostatic potential between the surface and the eluent, G is free energy of adsorption, and F is Faraday constant.

The equation states the retention of ionic solutes are governed by hydrophobic and ionic interaction. Using the theoretical models the retention and selectivity in ion pairing chromatography can be predicted to some extent. So many parameters affect it which are not independent. Recently the electrostatic theory of ion pairing chromatography is largely supported by experimental data and explains the retention in a wide range of experimental conditions; however, it is difficult to evaluate mathematically. The dynamic ion exchange and ion pair model show a good and easily understandable picture for practical evaluation of ion pair chromatography.

Apparatus

Isocratic chromatographic equipment with UV or electrochemical detector

Ultrasonic bath

Weighing machine

Graduated cylinder

Graduated flask, 25 cm³

Sperisorb ODS, 5 S, 150 × 4.6 mm, or other reversed-phase packing

pH meter

Chemicals

Vanillylmandelic acid (VMA)

Homovanillic acid (HVA)

5-Hydroxyindole acetic acid (5HIAA)

Epinephrine (adrenalin) (E)

Norepinephrine (noradrenalin) (NE)

Dopamine (DA)

3,4-Dihydroxyphenylacetic acid (DDPAC)

Stock solution of these compounds prepared in 0.1 M hydrochloride acid in a concentration of 100 μg/cm³

Working solution must be diluted from these solutions (concentration is about 1 μg/cm³)

Eluent: 0.065 M citric acid sodium acetate (pH = 4.3) containing 1.5 mM octane sulfonic acid sodium salt, 0.15 mM Na₂EDTA, and 7 v/v acetonitrile

Sodium sulfate

Procedure

> Prepare an eluent described with chemicals except 1.5 mM octane sulfonic acid, equilibrate the column, and inject the solution containing each of the solutes, flow rate is 1 cm^3/min, detector wavelength is 228 nm, or measuring potential 750 mV (Ag/AgCl).
>
> Change the eluent containing the ion paring reagent (1.5 mM octane sulfonic acid), inject sample from starting until no more change in retention time.
>
> Add to eluent 10, 50, and 100 mM Na$_2$SO$_4$ and repeat the measurement at least twice.
>
> Increase the acetonitrile concentration to 10 v/v and repeat the measurement after equilibration, at least twice.
>
> Decrease the acetonitrile concentration to 3 v/v and repeat the measurement after equilibration at least twice.
>
> Change the pH to 3 and 4.5 by adjusting the ratio of citric acid and sodium acetate and keep constant their overall concentration.
>
> Change the concentration of ion pairing reagent between 0.5 and   3.0 mM in the eluent.

Evaluation

Calculate all data (t_R, k, α, N, R_S) and plot k vs. concentration of ion pairing reagent in the eluent, k-pH, k-salt concentration in the eluent, and k-organic modifier concentration.

20.1.4. Evaluation of Column Efficiency and Selectivity in Ion Chromatography

Theory

The exact definition of ion chromatography is subject of controversy even today. The most widely accepted definition says every separation can be called ion chromatography where the separated solutes are in ionic form at the eluent condition used. The original technique developed by Small et al.[65] was mainly used for separation of inorganic anions and hydrophilic acids and bases. In the original framework a method for determination of ions was worked out: mobile phase ion chromatography[66] and ion exclusion chromatography.[67] One of the most common features of these methods is that the background conductivity is decreased by removing ions from eluent or transfer to a weakly dissociating compound. This technique is called eluent-suppressed ion chromatography. There is another approach where the same configuration is used as in conventional HPLC. No suppressor is needed for determination of ions.

Several terms are used to make the distinction from eluent-suppressed ion chromatography: "suppressorless" IC, "nonsuppressed" IC, and "electronic suppression", but the term "single column" IC (SCIC) has come to be generally accepted.[68]

The two types of ion chromatography are developing parallel, applying nearly all types of liquid chromatographic method and instrumentation. In spite of this fact there are some main features of ion chromatography. One is the column type used for most separations. In both eluent-suppressed and SCIC a low-capacity ion exchanger is used; the other one is the detection method applied. The "workhorse" detector is the conductivity detector. Depending on solutes, other detectors can also be used; for example, detectors can be used for UV absorbance both in direct and indirect absorbance measurement,[69-71] amperometric detection,[72-74] potentiometric detection with a metallic copper electrode[75] or with other sensors,[76] or refractive index detection,[77] via post column reactions.[78] The unique characteristic of ion chromatography is based on its selectivity. The chromatographic selectivity using an ion exchanger column mainly determines the stationary phase used. It means the main factor to alter the selectivity can be achieved by changing the column type. The selectivity of ion exchange column depends on the hydrophobic and charged or hydrophilic surface of the stationary phase. The stationary phase used in ion chromatography can be prepared from resin-based materials or silica gel. The resin-based packing prepared from styrol-divinylbenzene or acrylate resins. In eluent-suppressed ion chromatography for anion separation, the mixture of bicarbonate and carbonate in the millimolar range is generally used; in SCIC borate/gluconate and phthalate are popular.

In this chapter we deal with only the main application of ion chromatography. Today this chromatographic method is a routine technique for determination of inorganic anions.

Laboratory Equipment and Apparatus

Ion chromatographic systems should consist of the following components:

> Eluent reservoir
> Pump having a very low pulsation
> Anion exchange column
> Conductivity detector (with or without a suppressor device assembly) and/or UV detector
> Recording device
> Water purifying system

Additional pieces of equipment are drying oven, desiccator, graduated flasks with nominal capacity of 10, 100, and 1000 cm^3, graduated pipettes with a nominal capacity of 1, 10, and 10 cm^3, and membrane filtering apparatus with membrane filters of pore size 0.45 μm.

Reagents and Chemicals

In the test procedure use only reagents of specified, analytical grade chemicals. The water must have an electrical conductivity of 0.1 μs/cm. Eluent shall not contain particulate matter of a particle size 0.45 μm.

The solutes, eluents, and reference material are prepared from the following chemicals:

> Water S < 0.1 μs/cm
> Acetonitrile, chromatographic grade
> Methanol, chromatographic grade
> Glycerin
> n-Butanol, chromatographic grade
> Sodium hydrogen carbonate
> Potassium hydrogen phthalate
> Sodium tetraborate
> Boric acid
> D-Gluconic acid
> Sodium glucanate
> Lithium hydroxide monohydrate
> Potassium hydroxide
> Sodium bromide
> Sodium chloride
> Sodium nitrite
> Sodium nitrate
> Potassium dihydrogen phosphate
> Sodium sulfate

Prepare the eluent for nonsuppressed or single column ion chromatography. To test the efficiency and selectivity of the single column chromatography a borate/gluconate eluent can be prepared. For the preparation of borate/gluconate eluent first a borate/gluconate concentrate must be made. For this purpose take a 1-l volumetric flask and add 16 g sodium gluconate, 18 g boric acid, and 25 g sodium tetraborate decahydrate. Add approximately 500 cm^3 high purity water and mix thoroughly until dissolution of solids, then add 250 cm^3 of glycerin. Fill the flask to the mark with high purity (S < 0.1 μs/cm) water and mix thoroughly. The concentrate can be stored in the refrigerator up to 6 months.

Take another 1-l volumetric flask and pour 500 cm^3 high purity water, then add 20 cm^3 borate/gluconate concentrate, 20 cm^3 n-butanol, and 120 cm^3 acetonitrile. Fill the flask to the mark with high purity water and mix thoroughly. Before using this eluent, filter through a 0.45-μm membrane filter.

For a stock standard solution take a 1-l volumetric flask and weigh the quantity of substance specified in the following table.

Mass of Substance of Different Salts for Stock Solution

| | Anion substance concentration | |
	(g/dm³)	
Chloride	NaCl	1.6484
Bromide	NaBr	1.2877
Nitrite	NaNO₂	1.4998
Nitrate	NaNO₃	1.3707
Phosphate (ortho)	KH₂PO₄	1.4330
Sulfate	Na₂SO₄	1.4790

From this stock solution the working standard can be prepared. From the chloride, bromide, nitrite, and nitrate solution 1 cm³ and from the phosphate and sulfate solution 2 cm³ must be pipetted into a 100-cm³ metric flask. In this case the working standard concentration for Cl^-, Br^-, NO_2^-, and NO_3^- is 10 mg/dm³; for SO_4^{2-} and PO_4^{3-} the standard concentration is 20 mg/dm³. From the working standard 20 µl must be injected into the ion chromatographic equipment. Adjust the flow rate to 0.2, 0.5, and 1.0 cm³/min and inject the working standard solution at least twice.

Evaluation of Results

Determination of efficiency and selectivity can be done the same way as RP-HPLC and NP-HPLC.

For calculation of k the t_M must be determined. The t_M in single column ion chromatography is the retention time of cations. If a 10 mg/dm³ chloride solution is injected, the retention time of the cation will be used for calculation of the capacity ratio. If the ion exchange column is good, all pairs of anions will be separated. Calculate the N, α, R_S, H, and h. Plot the H vs. u and h vs. u.

Mobile Phase for Anion Exchanger on a Silica Gel Basis

In theory, we have emphasized the anion exchanger on silica gel base uses only eluents in a pH range of 1.5 to 6.5. To check the efficiency and selectivity this type of anion exchanger potassium hydrogen phthalate eluent must be used. For this procedure first an eluent concentrate must be prepared and this concentrate can be diluted. For preparing the eluent concentrate, place 20.5 g potassium hydrogen phthalate into a volumetric flask of nominal capacity of 1000 cm³, dissolve in high purity water, and make up to volume. The concentration of potassium hydrogen phthalate is 0.1 M, and can be used for a longer period if stored in a refrigerator. Take 20 cm³ of this concentrate and pour into a volumetric flask of a nominal capacity of 1000 cm³, add 100 cm³ methanol and dilute with high purity water to nearly 1000 cm³; place the solution into a beaker of nominal capacity of 1000 cm³ and adjust the pH to 5 with 0.1 M KOH solution. Place the solution again into the volumetric flask of nominal capacity of 1000 ml and make up to volume with high purity water. The potassium hydrogen phthalate concentration of the eluent is 2.1 mM and 10 v/v of methanol.

This eluent can be used for the determination of chloride, nitrate, phosphate, and sulfate in a single run. The test solute concentration is the same which was stated before, and calculation and presentation of data must be done the same as was given in the determination of efficiency and selectivity of the anion exchanger on a polymer base.

Mobile Phase for Ion Chromatography with Suppresser Technique

For the determination of anions by ion chromatography with suppressor technique sodium hydroxide and salt solution of weekly dissociated acids are used. Sodium carbonate/sodium hydrogen carbonate, sodium hydrogen carbonate, and sodium tetraborate are used. These days different suppression devices are known, but most of the application sodium carbonate/hydrogen carbonate is used as eluent. For determining the efficiency and selectivity if using an ion chromatographic system with suppressor technique, the eluent preparation is the following. Place 25.4 g sodium carbonate and 25.5 g sodium hydrogen carbonate into a volumetric flask of nominal capacity of 1000 cm^3, dissolve in high purity water. The concentration of sodium carbonate and sodium hydrogen carbonate are 0.24 and 0.3 M, respectively. This solution can be stored for several months in a refrigerator. For preparing the eluent, take 10 cm^3 of the concentrate into a 1000-cm^3 volumetric flask and make up to volume with high purity water.

For determination of efficiency and selectivity, the same solutes must be prepared as described above in the Reagents and Chemicals section. The eluent can be used for determination of fluoride, chloride, bromide, nitrite, nitrate, phosphate, and sulfate in a single run. Calculation and presentation of the results are the same as was given above.

20.1.5. Evaluation of Column Efficiency and Selectivity in Size-Exclusion (Gel-Permeation) Chromatography

Theory

The base of separation in size exclusion (SEC) or so-called gel-permeation (GPC) chromatography distribution of the solutes between the mobile phase and the stagnant portion of the eluent was retained within the pores of the stationary phase. Depending on the size of the molecules (in a given solution), they can or cannot penetrate into the pores, and hence the retention of molecules having different sizes (molecular weight) will be different. If the size of the solute is higher than the geometrical diameter of pores, it will excluded from it and travel through the column without any retention. The solute molecules remain in the mobile phase and their retention volume equals the interstitial volume of packing, called the exclusion volume. The small solute molecules can penetrate into the pore space of the packing without any barrier. Their retention volume will be equal to the interstitial or void volume and the volume of solvent held stationary in the pores of packing. This volume is called permeation volume. Stationary phases used in SEC can be characterized by two parameters: the exclusion and the permeation limit. In practice we present the dependence of the mean molecular weight of solute polymer on elution volume on a log-linear scale. The calibration curves of log M_W vs. V_E are valid only for the measured sample, eluent, and packing. From this curve the following information can be obtained:

Exclusion limit of the given packing, which refers to the mean molecular weight

Permeation limit of the given packing, which refers to the mean molecular weight

Working range of packing, which is the linear part of calibration curve, where selective permeation takes place

Molecular weight selectivity, which is expressed by the reciprocal of the slope of the calibration curve in linear range

The smaller the slope the higher is S and the better the selectivity of molecular weight (size) differences.

The capacity factor in SEC can be expressed as

$$k_{SEC} = \frac{V_A}{V_i}$$

where V_A is the available pore volume and V_i is the interstitial or void volume in the column.

This can be expressed by the effective pore volume (V_p)

$$k_{SEC} = K_{SEC} \frac{V_p}{V_i}$$

The retention volume can be given by the following equation:

$$V_R = V_i + K_{SEC} V_P$$

From the equation given above the K can be expressed:

$$K_{SEC} = \frac{V_R - V_i}{V_P}$$

The value of the distribution constant used in SEC at exclusion limit is 0 and is 1 at the permeation limit. The pore volume of stationary phase is not always known; an alternative definition is

$$K_{SEC} = \frac{V_R - V_i}{V_t - V_i}$$

where V_t is total volume of the column

To get molecular weight information from the measured value of elution volume we know the relationship between K_{SEC} and the molecular weight. The separation in SEC depends on the molecular sizes and not the molecular weight. The size of a molecule in a solution depends on the solvent used. Biopolymers have different conformation in different solvents (ionic strength, pH) and the retention as well.

According to Bly[79] the proposed separation mechanism[80] can be divided into three categories: steric exclusion,[81] restricted diffusion,[82-84] and thermodynamic theory.[85,86]

The retention of polymers is affected not only by the size of own but the size and shape of the pores of the packing material. For polymers different quantities are suggested to characterize the size and shape of molecules. The external length, $\overline{L}$, for rigid molecules,[87] the hydrodynamic volume (V_h)[88] the radius of gyration. The radius of gyration and hydrodynamic volume[89-91] are the basis of the universal calibration.[87] In this approach the hydrodynamic volume of a solute is equal to the product of the intrinsic viscosity (η) and the molecular weight of the polymer (M) plot of the logarithm of the $\eta \times M_w$ vs. the elution volume gives a universal calibration curve that can be used for the determination of molecular weight for all polymers. The calibration of the working range of the packing needs polymers of narrow molecular weight distribution. The universal calibration technique has been checked,[92-94] but one has to take into consideration that there is no direct relationship between the elution volume and (η) M. Universal calibration is one approach to get data for macromolecules; the other is the use of a molecular weight detector and a concentration detector in series,[90,91,95-98] provides data for calculation of absolute molecular weight. The polymers are a polydispersal system; therefore, in SEC it can be defined as a resolution factor which reflects the ability of the column to separate solutes of different molecular weight. The specific resolution factor (R_{SP}) is defined as

$$R_{SP} = R_S \left[1 / lg(M_1/M_2) \right]$$

where R_S is the usual term for resolution used in other branches of chromatography and M_1 and M_2 are the molecular weight for the calibration standards

The standards must be a narrow molecular distribution (polydispersity is about 1.1) and they differ about tenfold in average molecular weight. The minimum molecular weight ratio is a useful parameter for comparing different column performance (R_M is 1 or 1.5). If we consider the commonly used equation for R_S:

$$R_S = \frac{1}{4} N^{\frac{1}{2}} \frac{\alpha - 1}{\alpha} \frac{k}{k + 1}$$

the selectivity of SEC column is low, the capacity factor falls between 0 and 1, then the retardation term $\left(\frac{k}{k+1}\right)$ largely influences the resolution and achieves a high resolution the total number of theoretical plates should be high. For the theoretical plate in SEC is defined as:

$$N_i = \left[\frac{V_{E(i)}}{\sigma_{V(i)}}\right]^2_{u=const}$$

where $V_{E(i)}$ is the elution volume and $\sigma_{V(i)}$ is the total variance of an eluted peak, which is a sum of three independent contributions:

$$\sigma^2_{V(i)} = \sigma^2_{V(extra-column)} + \sigma^2_{V(column)} + \sigma^2_{V(MWD)}$$

where $\sigma^2_{V(extra-column)}$ is caused by instrumentation (injection volume, detector cell volume); $\sigma^2_{V(column)}$ is the dispersion on the chromatographic column caused by diffusion in the stagnant mobile phase and mass transfer resistance, the third-term of peak broadening source; longitudinal diffusion is small because polymers have low diffusivity; and $\sigma^2_{V(MWD)}$ is the variance of the peak due to the molecular weight distribution of polymer solutes.

The higher polydispersity of a sample, the higher its peak width. The only way to evaluate and compare the column used in SEC is to use a standard which has $p \leq 1.1$. In practice comparing column small molecules is used for determination of N and HETP such as toluene, acetone, benzyl alcohol, or ethyl benzene. These solutes can penetrate into the pores without any mass transfer resistance and therefore the measured values are higher than for macromolecules. The peak capacity (PC_{SEC}) is related to the plate number:

$$PC_{SEC} = 1 + 0.2 N_i^{\frac{1}{2}}$$

The value of PC_{SEC} is considerably smaller than can be obtained for RP-HPLC, NP-HPLC, and other liquid chromatographic methods used for separation of small molecules and typically is between 20 and 40.

The smaller analysis time in SEC is given by the retention time of the totally permeating solute used for determination of analysis time using a given column:

$$t_M = \frac{\eta L^2 \Phi}{d_p^2 \Delta p}$$

where η is the viscosity of the eluent used, L is the column length, Φ is the column resistance factor, d_p is average stationary phase diameter, and Δp is the pressure drop on the column.

One of the main parameters which influences the analysis time is the temperature. If the temperature increases the viscosity will decrease, and the pressure drop on the column will decrease. The increased temperature will increase the plate number by a factor of two or three, decreasing the mass transfer resistance. The retention is a temperature-independent process in SEC. If a temperature-dependent retention can be obtained, the solute can adsorb onto the surface of the support used in SEC. This mainly happens when biopolymers are separated with an inorganic support such as silica gel. To reduce this unfavorable effect surface modification or adding masking agent to the eluent are widely used. It is also found that separation is independent of flow rate. The SEC can be used for separation of both synthetic polymers and biopolymers. One class of the rigid and porous stationary phase is based on silica gel.[99] Unger stated the separation of polymers on silica gel columns can be done if the two solutes differ in their molecular weight by a factor of 2. There are two possibilities to separate a polydisperse solute. The first is a solution coupling of the columns of decreasing exclusion limit or using a packing which has two distinct average pore sizes with narrow distribution and nearly equal pore volume. This can be reached by mixing two different stationary phases or using a special stationary phase which meets the requirement stated above, a bimodal pore size which differs in diameter by of a factor 10.

Apparatus

Isocratic liquid chromatograph equipped with UV detector, UV wavelength is 215 nm for peptides and proteins and 254 nm for polystyrol standards

Volumetric cylinder, 1000 cm³

Weighing machine

Ultrasonic bath

Column: 300 × 4 mm, packing LiChrospher Si 100 and LiChrospher Si 500, d_p = 10 μm

TSK gel SWG 2000 and TSK-SWG 3000 (250 × 4.6 mm)

Chemicals

Protein standards:
1. Cytochrome C, M = 12,500
2. Trypsin inhibitor, M = 24,000
3. Ovalbumin, M = 45,000
4. Bovine serum albumin, M = 68,000
5. Glucose product M = 135,000
6. Gammaglobulin, M = 167,000
7. Catalase, M = 240,000
8. Ferritin monomer, M = 450,000
9. Blue dextran, M > 10^6
   0.1 M KH_2PO_4 pH = 7.0 and 0.3 M NaCl
   Polystyrene standards: between mol weight 2000 and 2.2 × 10^6
   tetrahydrofuran
   ethyl benzene

Procedure

*Evaluation of SEC Efficiency and Selectivity for Separation of
Peptides and Proteins*

For evaluation of efficiency and selectivity in SEC a silica gel-based column
must be used. We suggest TSK gel SW G-2000 and TSK gel SW 3000 (Toyo
Soda, Japan) which is supplied by BioRad under the name of BioSil TSK 125
and BioSil TSK 250, respectively. The working range is between 500 and
60,000 to BioSil TSK 125 (TSK G-2000) and between 1000 and 300,000 for
BioSil TSK 250 (TSK G-3000). The column is stored under 10/90 v/v
methanol/water. Prepare the eluent weighing the calculated amount of KH_2PO_4
and NaCl for preparing 1 dm^3 eluent. Filter and degas it. Purge the instrument
about 5 min with eluent and let it run about 30 min with a flow rate of 1 cm^3/min.
The sample concentration must be between 0.3 and 1.0 mg, the peptide solute in
0.1 M KH_2PO_4 pH = 6.8 and protein in 0.1 M KH_2PO_4 pH = 7 and 0.3 M NaCl;
add 9 mg/cm^3 sodium aside as an antimicrobial compound during storage.

　　　Inject from every standard solution twice for each column; inject acetone
twice. After finishing the measurement first wash the column with water then
with 10/90 v/v methanol/water mixture.

Evaluation

Construct a plot log M_W vs. V_E for the two columns. Give the working range
of the two columns, calculate the slope of the linear range of calibration curve,
and give "S" (molecular weight selectivity). Calculate the N_{SEC}, k_{SEC}, and R_{SP}.

*Evaluation of SEC Efficiency and Selectivity for Separation of
Synthetic Polymers*

Take a rigid silica gel or modified silica gel column such as LiChrospher 100
Å and LiChrospher 500 Å and couple them. Wash them with tetrahydrofuran

at a flow rate of 1 cm^3/min. Prepare the polystyrol standard molecular weight range from 2000 to 10^6 solved in eluent (0.1 to 1 mg/cm^3) and inject into the column coupled at least twice. For determination of permeation limit, inject ethyl benzene again at least twice.

Evaluation

Evaluation is written the same as for the evaluation of SEC efficiency and selectivity for separation of peptides and proteins.

20.2. Summary of Chromatographic Quantities

Equations for Calculation of Chromatographic Data

Connected Retention Time or Netto Retention Time (t_N)

$$t_N = t_R - t_M$$

where t_R = retention time of a compound measured (min)
t_M = retention time of unretained compound (min)

Capacity Factor (k)

$$k = \frac{t_R - t_M}{t_M} = \frac{t_N}{t_M}$$

Retention Volume (V_R)

$$V_R = t_R F$$

Netto Retention Volume (V_N)

$$V_N = t_N F$$

where F = flow rate [cm^3/min]

Relative Retention or Separation Factor (α)

$$\alpha = \frac{k_2}{k_1}$$

where $k_2 > k_1$

Resolution (R_s)

$$R_S = 2 \frac{t_{R_2} - t_{R_1}}{w_1 + w_2}$$

where w_1 and w_2 = band width at baseline, t_R and w must be the same dimension

$$R_S = \frac{1}{4}\sqrt{N}\ \frac{\alpha-1}{\alpha}\ \frac{k}{k+1}$$

where $k = k_2$

$\quad\quad N = N_2$

Number of Theoretical Plate or Plate Number (N)

$$N = 16\left(\frac{t_R}{w}\right)^2 = 5.54\left(\frac{t_R}{w_{\frac{1}{2}}}\right)^2$$

where w and $w_{\frac{1}{2}}$ peak width at baseline and at half height, t_R and w must be the same dimension

Effective Plate Number (n):

$$n = N\left(\frac{k}{1+k}\right)^2$$

System or True Plate Number (N_{sys})

$$N_{sys} = \frac{41.7\left(\dfrac{t_R}{w_{0.1}}\right)^2}{\dfrac{b}{a}+1.25}$$

where $w_{0.1}$ = peak width at 0.1 height

$\quad\quad w_{0.1} = a + b$ and are defined as indicated in Figure 51.

Plate Height or Height Equivalent of Theoretical Plate (H or HETP)

$$H = \frac{L}{N}$$

where L = column length (μm)

Reduced Plate Height (h)

$$h = \frac{H}{d_p}$$

where d_p = average particle diameter (μm)

Linear Velocity (u)

$$u = \frac{L}{t_M}$$

Asymmetry Factor (δ)

$$\delta = \frac{b}{a}$$

where a and b are defined as indicated in Figure 20-1.

Resistance Factor (Φ)

$$\Phi = \frac{\Delta p \, \bar{d}_p^2 \, t_M}{\eta L^2}$$

where Δp = column pressure drop (Pa)
   $\bar{d}_p$ = average particle diameter (m)
   t_M = retention time of unretained compound (s)
   L = column length (m)

The Chromatographic Permeability (K)

$$K = \frac{\bar{d}_p^2}{E}$$

Separation Impedance (E)

$$E = h^2 \Phi = \frac{H^2}{K}$$

20.3. Some Features of Isocratic vs. Gradient Elution Techniques

Theory

During most of the separations, eluent composition is held constant. This elution technique is called isocratic (isocratic = equal strength). In gradient elution the mobile phase composition is not constant, but gradually changed in time (solvent programming). The solvent strength is increased during the run and strongly retained solutes will elute earlier.

By using gradient elution the general elution problem can be solved: if a sample contains solutes with wide polarity range, the early eluting peaks have retention times near the unretained compound and of course are poorly resolved, the late eluting solutes can hardly be distinguished from baseline noise. Gradient elution offers the solution for the problem mentioned above. To start a weak eluent composition the early eluting peaks can be resolved; at higher elution strength the retention time of the late eluting compound will decrease dramatically and the concentration of solute at peak maximum will increase, that is, the sensitivity of the chromatographic system will increase.

The most commonly used approach in gradient elution is the linear solvent strength theory (LSS).[100-103] The LSS approach means the retention of given solutes decreases exponentially with gradient run time. If the gradient conditions are chosen correctly, the following experimental advantages can be given:

Effective average capacity factor is roughly equal for all solutes.
Approximately equal peak widths for each solutes.
All components have the same concentration at peak maximum; the sensitivity does not depend on retention time.
Reliable resolution for all components and no bunching in chromatogram compared to isocratic separation.
The resolution between adjacent band pairs with similar values of separation factor are equal.

The retention (capacity factor) during the run can be derived according to

$$\log k_i = \log k_0 - b\left(\frac{t}{t_M}\right)$$

where k_0 is the value at $t = 0$ for the solute interested at the beginning of gradient run, k_0 is equal to the value of the capacity factor with the weaker solvent used in gradient, "b" is the gradient steepness parameter, t is the time after the start of the gradient and sample injection, and t_M is the column holdup time or retention of unretained compound.

If the value of "b" is constant and has the same value for all sample components and the value of capacity factor is in the optimum range $1 < k_i < 10$, then the equation can be used for the prediction of the capacity factor for all components measured. It has already been shown that the predicted and measured values fit reasonably well. However, in practice the gradient steepness parameter will vary with the solvent strength; therefore, it is only roughly constant for low molecular weight and structurally similar solutes.[104-109] For gradient steepness parameters there are several expressions:[101]

$$b = t_M \Phi' \lg \frac{k_a}{k_b} = \frac{t_M}{t_g} \lg \frac{k_a}{k_b}$$

where t_a is the column holdup time (retention time of an unretained component, equal to t_M), Φ' is the rate of change with time of solvent composition is referred to as gradient steepness (change in volume fraction B/time), k_a is the retention time at starting eluent (k_a is practically equal to k_0), and k_b is retention factor at final stage of gradient elution.[111]

$$b = \Delta \Phi S \frac{t_M}{t_g}$$

where $\Delta \Phi$ is the change in the volume fraction of the stronger eluting solvent during the gradient and t_g is the time from the start to the end of the gradient, sample is injected when the gradient elution starts.

The equations can be easily used for any calculation if linear solvent strength is applied and S is fairly constant. For this situation equations used in isocratic elution can be used. In reversed-phase chromatography the variation of retention can be approximated by

$$\ln k = \ln k_w - \varphi S$$

where k_w is the solute capacity factor with water eluent

At gradient elution instead of k_w we use k_a and k_b for k and express the φS from equation expressed above:

$$\ln \frac{k_a}{k_b} = \varphi S$$

If we introduce φS to the original Snyder equation we will get the equation given by Poole. In his work only the difference $\Delta \Phi$ must be used instead of simple solvent volume fraction of stronger eluent component (φ) in the mobile phase.

In RP-HPLC the linear solvent strength gradients are the most popular and the model reasonably accurate to predict retention and resolution. A number of experiments is enough for computer simulation.[112-115] In normal phase chromatography the situation is a little bit different. The selection of A and B solvents must be taken carefully to avoid solvent demixing. Usually for A the pure hexane or other apolaric solvent must not be chosen.

It helps to add 0.1 v/v organic modifier to hexane and solvent demixing can be avoided. In most of the cases concave gradient shape is recommended.[116]

The gradient elution technique vs. isocratic separation and the relationship between the two chromatogram development modes need to know what is the relationship between the parameters which describe the retention, resolution and kinetic efficiency. It can be shown there is a good correlation between the gradient steepness parameter and capacity factor in isocratic separation. Snyder has already shown this comparing the retention in isocratic and gradient elution:

$$t_R - t_M = t_M k \qquad\qquad \text{for isocratic elution}$$

$$t_g - t_M = t_M \left[\frac{1}{b} \lg(2.3\,bk_0 + 1) \right] \qquad \text{for gradient elution}$$

The equation is valid for all bonds that elute a time $t_g < t_G + t_M$. In gradient elution k_0 (or k_A) is usually $\gg 1$, and $b < k$, then equation for gradient elution becomes

$$t_g - t_M = t_M \left(\frac{1}{b} \lg k_0 \right)$$

Comparing the equations we can conclude the inverse of gradient steepness parameter $(\frac{1}{b})$ has a similar effect on a separation as does a change in the capacity factor for isocratic separation. If we express the apparent capacity factor in gradient elution:

$$k_g = \frac{t_g - t_M}{t_M} = \frac{1}{b} \lg k_0$$

At a given concentration of mobile phase the apparent capacity factor can be changed by changing the value of gradient steepness parameter.

In gradient elution the final capacity factor (k_f) can be calculated as (this k can be calculated from the chromatogram the same way that is done in isocratic separation):

$$k_f = \frac{1}{2.3b + \dfrac{1}{k_0}}$$

If k_0 (k_b) is large, it is always true for later eluting solutes:

$$k_f = \frac{1}{2.3b}$$

In isocratic separation the capacity factor depends on distribution constant (D) as

$$k = \Phi D$$

where Φ is the phase ratio.

Comparing the two equations $\frac{1}{b}$ and D have the same effect on retention. In isocratic elution D can be changed by mobile phase composition and k_f depends on $\frac{1}{b}$. In isocratic separation the optimum value of k varies between 2 and 5, the $\frac{1}{b}$ value in gradient elution is between 4 and 9. Higher the value of b, less the analysis time. In chromatography the band width is determined by the time the solute spends on the column. The higher the retention, the higher the peak width. In gradient elution there is an additional effect which is caused by band compression that reduces the band width by factor G.

The band width in isocratic and gradient elution can be compared:

$$t_w = \frac{4}{N^{\frac{1}{2}}} t_0 (1 + k) \qquad \text{for isocratic elution}$$

$$t_{wG} = \frac{4}{N^{\frac{1}{2}}} t_M (1 + k_f) G \qquad \text{for gradient elution}$$

If we express k_f by b and substitute to t_{wG} equation:

$$t_{wG} = \frac{4}{N^{\frac{1}{2}}} t_M \left(1 + \frac{1}{2.3b}\right) G$$

This equation is true if the late eluting peak or k_0 (k_w) is high. In gradient elution the peak width is determined by k_f, G, and normal band broadening.[100,117-119] The $\sigma_t = (2.3b + 1)Gt_M/2.3bN^{1/2}$ for the late eluting solutes and $\sigma_t = G(1 + k_0/(2.3k_0b + 1)t_M/N^{1/2})$ for early eluting bonds.

The bond compression phenomenon arises from faster migration of the tail of bands. In isocratic elution the two migration rates are equal. For the usual values of b the value of G is close to unity.

The resolution between two adjacent solutes can be expressed by a similar expression that is used in isocratic separation:

$$R_S = \frac{1}{4} N^{\frac{1}{2}} \frac{\alpha - 1}{\alpha} Q$$

The quantity of Q can be expressed as a function of b:

$$Q = \frac{1}{1.15b} \Big/ (1.15b + 1)$$

where $\dfrac{1}{1.15b}$ is equal to medium value of k (k_c) when the band is in the midpoint of the column and

$$R_S = \frac{1}{4} N^{\frac{1}{2}} \frac{\alpha - 1}{\alpha} \frac{\dfrac{1}{1.15b}}{\dfrac{1}{1.15b} + 1}$$

Again $\frac{1}{b}$ has the same effect on resolution as k in isocratic separation. One of the main advantages of gradient elution arises from near equal band width, because the detection limit is inversely proportional to σ_t. If value of b is high the bands can be compressed as to give narrower bands than those via isocratic elution at k = 0.

In practical application one main question is the gradient shape. This depends on the relationship between log k vs. ϕ_b. In RP-HPLC log k vs. Φ_b can be treated as linear for binary eluent. Most of the practical applications can be done using binary gradient in which the volume fraction of the stronger eluting solvent is progressively increased by time. Ternary and quaternary solvent gradients provide a better control of selectivity but are difficult to program and interpret.

In normal phase chromatography log k vs. Φ shows a convex curve; therefore, concave curves are to be preferred.

In ion-exchange chromatography when salt gradient is used, concave gradients are to be preferred; at pH gradients linear gradients can be used.

In ion pair chromatography the situation is even more complicated. There are several possibilities for changing the selectivity and retention. For example, we can alter the eluent composition, concentration if ion-pairing reagent, pH, and concentration of salt concentration. If log k vs. parameter changed is linear, linear gradients are to be preferred; if log k' vs. parameter changed is convex, concave gradient preferred. If log k' vs. parameter changed concave, convex gradient preferred.

The mathematical description of these gradients as a function of time (t) and total gradient time (t_G) is

$$\text{linear gradient}: \quad \Phi_b = \frac{t}{t_G}$$

$$\text{convex gradient}: \quad \Phi_b = 1 - \left(1 - \frac{t}{t_G}\right)^n$$

$$\text{concave gradient}: \quad \Phi_b = \left(\frac{t}{t_G}\right)^n$$

where Φ_b is the volume fraction of stronger eluent component, $n = 1, 2, \ldots$

The former gradient can be classified into two main groups: high-pressure devices and low-pressure devices, according to mixing of the solvents used in gradient elution.

There are some general requirements for the gradient former:

Good mixing needs reproducible separation, precise quantitation, and low detection limit.

Good gradient reproducibility from 0 to 100% B in eluent. At the beginning and the end the reproducibility is critical, so if possible the gradient range must be restricted to the range 10 to 90% B in eluent.

Dwell time between the gradient mixer and column inlet must be minimal; otherwise, distortion of the gradient shape and slow solvent changeover will occur.

Apparatus

Gradient separation equipped with multichannel UV and fluo-
rescent detector
Ultrasonic bath
Weighing machine
Graduated flask, 25 cm^3
Pipettes
Bakerbond PAH 16-plus, 3.0 × 250 mm, or other PAH column
from Supelco, WATERS, Whatman, etc.

Chemicals

Standards for PAH: naphthalene, acenaphthylene, acenaphthene,
fluorene, phenanthrene, anthracene, fluoranthene, pyrene,
benzo-(a)-anthracene, chrysene, benzo-(b)-fluoranthene, benzo-
(k)-fluoranthene, benzo-(a)-pyrene, dibenzo-(a,h)-anthracene,
benzo-(ghi)-perylene, indeno-(1,2,3,cd)-pyrene; for separation
dilute to 1 µg/ml with acetonitrile/water 80/20 v/v
Acetonitrile;
Water;
Eluent for gradient elution: eluent A: 50 v/v acetonitrile/water,
eluent B: acetonitrile

Procedure

The Study of Shape of Gradient for Separation of PAH

Prepare the eluents and couple the multiwavelength and fluorescent detector.
Adjust the wavelength range of multichannel detector for measurement from 200
to 450 nm and fluorescence detector $\lambda_{EX} = 240$ $\lambda_{EM} = 400$. Start the gradient run
with a gradient time of 40 min and repeat it with the gradient elution with convex
and concave gradients n chosen freely. Repeat the gradient runs with a gradient
time of 30 and 50 min. Finally make a gradient run with the following parameters:

time (min)	vol. % of AN in the eluent
0–5	50% (eluent A)
5–35	50–100 (linear)
35–45	100 (eluent B)

Repeat the runs at least five times. Make an isocratic run at 80/20 v/v AN/water.

Evaluation

Calculate the t_R, α, and N at different gradient shapes and times. Make an
evaluation for gradient shape, gradient steepness parameters, and give the best

gradient elution for the determination of polycyclic aromatic hydrocarbons. Compare the signals of the two detectors at different selected wavelengths (UV). Compare the N at gradient run and at isocratic run for PAH. Calculate the average retention times and their standard deviation from the last five runs.

20.4. Analytical Application of HPLC

20.4.1. Analytical Application of HPLC in Synthetic Chemistry

Theory

HPLC is a separation technique which can determine the amount of a compound or impurities in a complex mixture. In analytical chromatography we need information regarding the composition of the mixture. The column is the heart of the separation system, but additional information can be obtained from the detector signal. Using a diode array detector (DAD), we can get a UV spectrum from the solutes detected. This additional information will confirm the quality of products. Later in this book we deal with these and other important detection techniques. We will pay some attention to the chromatographic method chosen for a given purpose and some criteria which must be applied for these selections.

To solve a particular problem the first step is the column selection. The aim of column selection is to select a stationary phase which gives the best possibility for separation and identification of a given compound. In the second stage of phase system optimization, the mobile phase composition must be selected to give a rugged method. The third step is the selection of the detection method. In the past it was believed that LC is a separation technique where only a few stationary phases can be chosen and good separation can be achieved by altering mobile phase composition. This ideal picture is over; each stationary phase has its own physical-chemical character, and column to column will change the retention, efficiency, and selectivity.

HPLC is an applied separation technique (science) with its own roles which have to be known; however, great success of this method lies in its practical application. For example, theoretical treatment of reversed-phase chromatography more or less is similar but if we go into details the term covers different classes of stationary phases: copolymer based, graphitized carbon, even silica gel or alumina or dynamically coated silica gel or alumina. The term octadecyl silica means an octadecyl group covalently bound on the surface of the silica, but chemically identical silica possesses different stability and selectivity, depending on row material, bonding chemistry, etc. The large number of commercially available reversed-phase stationary phases (about

200) will cause a difficult selection, because only a few data can be known about the surface and their physical-chemical properties.

There are several guidelines which try to simplify the method development on some basic criteria such as solubility of solutes, molecular weight, polarity, and so on. At the first stage of the method development it can serve as an empirical help, but for a more precise and applicable method development a sophisticated computer program-based expert system is needed. Computer programs for optimization are usually based on the molecular behavior of a given solute in the mobile phase. There are only a few indications about the stationary phase effects. Retention and selectivity in turn will determine of stationary and mobile phase and cannot be treated separately when a separation problem must be solved.

Now we try to show a practical application of how we can solve a relatively simple separation problem. A synthetic mixture contains 2,4-dichlorophenol as a main component and mono- and trichlorophenols (and polychlorophenols) as impurities.

Determination of the Main Components

Theory

Selection of Instrumental Conditions.   2,4-Dichlorophenol is a UV-absorbing compound and a single- or multichannel (photo diode array) detector can be used. For analytical determination there are no specific requirements, so a conventional analytical column can be used, 250×4.6 mm or 150×4.6 mm. Particle size of stationary phase is about 5 or 10 μm. Different isomers might be in the sample and polarity of compounds differ greatly; a gradient equipment must be chosen.

Phase Selection.   Hydrophobicity of the main component and its impurities differ to some extent. If we want to determine the main component only, we need to separate the main component and its impurities, but we are not interested in the separation of isomer impurities, we can choose reversed-phase chromatography. In reversed-phase chromatography (RP-HPLC) the octadecyl modified silica are used widely; we choose a C-18 or ODS phase. In our cases a Nova-Pak C-18, 4-μm, 100×3.9 mm cartridge will be used.

Organic solvents used in RP-HPLC are mainly limited to three solvents: methanol, acetonitrile, and tetrahydrofuran. Acetonitrile has UV cutoff at low wavelength and low viscosity; therefore, it can be used at a higher flow rate and detection, or detection wavelength range for diode array detector can be started at as low a wavelength as 190 nm.

Among phenols mainly the chlorophenols possess the acidic group, which can ionize in aqueous medium to get high kinetic efficiency ionization must be suppressed by adjusting the pH below the pK_a with 2 or 3 units. It means pH < 3 ionization of 2,4-dichlorophenol is suppressed. The pH can be adjusted by adding buffer to the eluent. Buffer must be transparent and soluble in eluent used. Phosphate buffers are widely used for adjusting the eluent pH used. Concentration of buffer used must be as low as possible, usually between 0.01 and 0.1 M.

Composition of partly aqueous solvent can be determined by trial and error based on an empirical knowledge of chromatographers or using a computer-based expert system[120,121] or from a gradient run. The theory of gradient run has been given in Section 20.3. The procedure for predicting the mobile phase composition of the isocratic separation from the gradient elution is the following:

1. Run an initial gradient from 100% buffer to 100% acetonitrile. The slope of the gradient is set at 0.2.

2. Calculate the k value (k_f value of a solute leaves as it leaves the column). The dichlorophenol is a relatively polar compound and in this case:

$$k_f = \frac{1}{2.3b} = 2.2$$

3. Calculate the solvent composition at the exit of the column at the time the peak of interest is eluted (Φ_e):

$$\Phi_e(AN) = \Phi'\left(t_g - t_M - t_d\right)$$

where Φ' is the rate of change with time of solvent composition, t_g is the retention time of solute during the gradient, t_M is the retention time of inert solute (dead time), and t_d is the delay time of the gradient system.

4. If the desired k value for dichlorophenol is 5, then the volume fraction of organic solvent is

$$\Phi_2 = \frac{1}{s(\lg k_f - \lg k)} + \Phi_e$$

where s = 5 (slope of lgk-eluent composition at isocratic elution)
$\Phi' = 0.04$
$t_g = 25$
$t_M = 2$
$t_d = 0.5$

In our cases first trial must be taken at about 60% acetonitrile.

5. If the $1 < k < 10$ then the organic modifier concentration can be adjusted to the desired capacity factor.

For validation of the method the following steps must be taken: linearity and range, limit of detection, limit of quantitation, recovery, day in reproducibility, and day to day reproducibility. To test the ruggedness of separation change the eluent composition with the value ±5% measured in point 5 and pH ±0.2 and calculate the capacity factor and resolution.

For determination of impurities and degradation products the detector response for 2,4-dichlorophenol is out of the linear range. In this case the concentration of solution injected onto the chromatographic column must be at least one order higher than it was for the determination of main components. The column in this case operates in a mass overload condition for 2,4-dichlorophenol. A column is considered to be overloaded when the solute capacity factor values change by more than 10% as the sample size increased. In concentration overload, the brand profile broadens and becomes asymmetric and the resolution will decrease for the closely eluted components. In analytical separation without concentration overload $R_S \geq 1.5$ means a baseline separation. In concentration overload the value of the R_S must be higher in pharmaceutical application is advisable if $R_S > 4$. If the separation between the two adjacent impurities of 2,4-dichlorophenol is less than 4 adjusting with the eluent composition must reach a value of 4.

Using a semi-preparative chromatographic methods is our aim to get enough pure solute for structural analytical methods such as mass spectrometry, infrared spectroscopy, and NMR. The amount of sample required dictates the size of the column. The simplest approach is to scale up an analytical separation used before for determination of given solutes. Increasing column dimensions will increase the amount of stationary phase in it and permit the use of higher column loading in the linear range of adsorption isotherm. If the sample loading is within the linear range of adsorption isotherm results obtained in an analytical separation can easily be extrapolated to a high-performance semipreparative separation.

The largest column size that can be used with an analytical instrument is about 25×2.5 cm and contains about 60 g silica-based stationary phase and an optimum flow rate of around 10 cm^3/min. To increase the column size (length and column inner diameter) to get higher loadability special instrumentation can be applied. Analytical columns are usually packed by pumping a slurry of a packing material at a pressure of up to 400 to 1000 bar. It is difficult to pack a large-size column in this manner with a small particle size stationary phase (5 and 10 μm). The recent tendency is to get as high efficiency as possible using a column for preparative purpose; 10 to 25 μm silica or modified silica are used to avoid reducing the efficiency of column compressed bed techniques are employed. Dry packing or slurry is placed in the column, which is then compressed. To avoid decreasing efficiency of a preparative column radial compression was developed.

A flexible walled cartridge filled with the packing material is placed in a chamber. At two ends of the column the cartridge was sealed and in the space between the cartridge wall and column wall pressurized with nitrogen or it can be done with liquid, too. In preparative liquid chromatography silica gel is widely used, but any stationary phase used in analytical chromatography can be used in preparative chromatography. The silica gel is relatively inexpensive and eluents used with silica gel have low boiling points. To recover the analytes from a dilute solution can be achieved at low temperature without destroying the thermal labile compounds.

The preparative chromatography can be classified according to column sizes, but any classification shows close connection to the amount of the material gained in one chromatographic run.

If the column diameter falls between 1 and 5 mm, then the column is the same as used in analytical chromatography; if $6 < d_c < 11$ m then the name of the preparative column is wide bore; if $10 < d_c < 30$, it is called long narrow; if $20 < d_c < 100$ the preparative column is called short thick; if $100 < d_c < 1000$, it is called industrial process scale preparative column. Loadability of columns will increase parallel with their diameter. While using a wide bore column 3 to 12 mg pure material can be gained, until using an industrial column as high as a kilogram range pure substances can be prepared. In analytical practice we usually need 1 g; therefore, a long narrow column is enough.

The success of preparative separation will be judged by the production rate and recovery yield obtained;[122,123] two reviews about very large scale chromatography and all aspects of preparative chromatography can found in References 123 and 124.

Liquid chromatography is a separation technique of small and high molecular weight compounds. To gain quantitative information from the chromatograms is relatively easy, but to get direct qualitative information from an unknown compound has not been solved yet. Using multichannel detection gives information about the UV or UV-VIS spectra; to get more information from chemical structure from solutes investigated, a mass spectrometer (MS) must be coupled directly to the chromatographic column. Interfacing HPLC to MS is not as easy as coupling GC to MS. Eluent must be vaporizing and creating a huge amount of vapor, and ionization efficiency of MS varies with the conditions used and for some compounds is relatively low. The principal method of interfacing HPLC to MS are

Direct liquid introduction
Continuous-flow fast atom bombardment
Moving belt
Thermospray
Electrospray
Various particle beam approaches

The thermospray interface is one of the most popular ones in use today. Fully compatible with eluent flow rate used today (1 to 2 cm^3/min), this interface is commercially available for most of the common quadrupole and magnetic sector mass spectrometers. It consists of a probe, source, and vacuum system combined in a single unit. The vaporizer probe contains a resistively heated capillary tube, which vaporizes the liquid and produces a high speed vapor containing very fine droplets and particles. From the ion source the generated ions are transmitted to the mass analyzer through a sampling cone placed in the path of the vapor jet. In addition to thermospray ionization an additional filament or discharge device is provided to assist in the ionization of samples introduced with nonaqueous or partly aqueous mobile phase. A vacuum system connected to the ion source will remove the excess vapor. The degree of vaporization of the mobile phase is a critical experimental parameter. Unfortunately, it depends largely on HPLC conditions used (eluent composition, flow rate) and ion source conditions.

One of the latest developments is the atmospheric pressure chemical ionization source. To this source the eluent from the column can be introduced directly. Unfortunately, no single LC-MS interface can be applied in a wide range of compounds and conditions used in HPLC. Several different approaches show that LC-MS is not at the stage where GC-MS works today.

Determination of Impurities in 2,4-Dichlorophenol Product

Apparatus

> Gradient elution system with photodiode array detector
> Ultrasonic bath
> Weighing machine
> Graduated flask, 25 cm^3
> 250×10 mm column packed with reversed-phase material (RP-18 or ODS $d_p = 10$ μm)
> Separator funnel 10 and 25 cm^3

Chemicals

The chemicals are the same as described for the determination of 2,4-dichlorophenol and saturated NaCl solution.

Procedure

Make a solution in acetonitrile, concentration of about 0.1 to 0.3 g/cm^3. Adjust the chromatographic conditions until $R_S > 4$ for the components will be prepared and its adjacent chromatographic peaks are in analytical separation. The separation criteria are about the same as the separation of the main component and adjacent impurities. Inject 200 μl and collect the eluent at the chromatographic peak interested. Repeat the cycles three or five times. Add a double amount of hexane to the collected sample, and add an equal amount of sodium chloride in a separator funnel and shake it. Remove the aqueous phase and extract the aqueous layer at least twice. Join the organic phase (extract) and in rotary evaporator evaporate it about 1 cm^3. Pour it into a conical tube and evaporate to dry with nitrogen at ambient temperature, except 50 μl which evaporates separately, and dissolve in the eluent and inject onto the chromatographic column for controlling the purity of prepared product. For each impurity this procedure must be done. The collected samples are pure enough if the impurities are not higher than 2%; the sample is ready for MS, IR, and NMR analyses.

Evaluation

Measure the sample collected and calculate the yield. From the chromatogram estimate the purity of collected sample.

Determination of 2,4-Dichlorophenol Content and Impurities by HPLC

Apparatus

>Gradient elution system with photodiode array detector
>Ultrasonic bath
>Weighing machine
>Graduated flask, 25 cm^3
>Column: 100×3.9 mm Novapack C-18, 4 μm
>Separator funnel

Chemicals

>2,4-Dichloro samples from different manufacturers
>Monochlorophenols
>2,4,6-Trichlorophenol
>Acetonitrile
>Water
>0.01 M KH$_2$PO$_4$ buffer pH = 3, adjusted with H$_3$PO$_4$

Procedure

Follow the instructions found in the theoretical part of the determination of the main components. Determine the isocratic mobile phase composition from one gradient run. Prepare stock solution from chlorophenols: 1.0 mg/cm^3 in acetonitrile, and make a working solution; dilute the stock solution with the eluent. Prepare a mixture of working solution containing about 10 μg/cm^3. Inject the sample onto the chromatographic column. If the separation $R_S \leq 1.5$, adjust the mobile phase composition to get about this value for R_S. If the capacity factor of 2,4-dichlorophenyl is higher than 6, adjust the mobile phase composition to get about this value.

For determination of mono- and 2,4,6-trichlorophenol impurities of 2,4-dichloro samples, adjust the eluent composition to get the R_S value of about 4 between the two adjacent impurities of 2,4-dichlorophenol. Prepare a solution from 2,4-dichlorophenol containing 1 to 10 mg/cm^3 in eluent; inject onto the chromatographic column. Follow the validation procedure given in the preceding section.

Evaluation

Give all parameters which are important for validation: limit of detection, limit of quantitation, linearity, day-to-day reproducibility, day-in reproducibility, recovery, retention data, concentration of the main component (2,4-dichlorophenol), and concentration of monochlorophenols and 2,4,6-trichlorophenol.

20.4.2. Application of HPLC in Environmental Analysis

Determination of Anions in High-Purity Waters

Theory

The typical limit of detection (LOD) in ion chromatography published by Johnson[125] for eluent-suppressed ion chromatography is between 5 and 10 ppb for weekly retained anions and 20 to 1000 ppb for the strongly retained one injecting 50 µl to the column. The limit of quantitation is about at least five times higher, and if we take into consideration what Johnson stated,[125] these values can reach — if the chromatographic conditions are ideal — the limit of quantitation for a real sample of about 0.1 ppm for weekly retained solutes.

Determination of anions at low level needs preconcentration. A short cartridge packed with same type of material used in separator column is attached to the injection valve in place of the sample loop. The sample is forced through this cartridge, the analyte will be sorbed onto the anion exchanger, and the retained anions will be desorbed by eluent, when the injection valve is turned into the injection position.

Apparatus

Single column ion chromatograph or an ion chromatograph with suppressor.

Anion concentrator cartridge
Ion exchanger/IC-PAK A HR (Waters, Milford, U.S.)
Trace enrichment pump
Ultrasonic bath
Volumetric flask
Graduated cylinder
Automatic pipettes
MilliQ water purification system

Chemicals

Borate/gluconate eluent

To a 1-l volumetric flask add: 16 g sodium gluconate, 18 g boric acid, 25 g sodium tetraborate decahydrate

Add 500 cm^3 high purity water, dissolve it, then fill the flask to the mark with MilliQ or other high purity water.

Take 20 cm^3 borate/gluconate concentrate and put into a 1-1 volumetric flask which contains 500 ml MilliQ water or other high purity water; add 20 cm^3 n-butanol, 5 cm^3 glycerin, and 120 cm^3 acetonitrile to the volumetric flask. Fill the flask to the mark with MilliQ water or other high purity water. The eluent must be filtered through a 0.22 or 0.45 μm Durapore membrane (GVWP).

The standard is prepared as described in Section 20.1.4. Silica-based ion exchanger or eluent suppressed ion chromatography is used; follow the instructions as described in the section.

Procedure

The eluent must be degassed before it is used. It can be done by placing into an ultrasonic bath or vacuum or helium purge. Purge pump about 5 min with eluent and adjust the flow rate to 1 cm^3/min. The thermal and chemical equilibrium take places between 1 and 2 hours. During this time the sample can be forced through the concentrator cartridge by a pump or by a syringe. After no baseline drift of conductivity detector at high sensitivity the sample valve must be turned to the injection position. After a few minutes turn back the sample valve to the load position and wash the column with high purity water, about 20 cm^3. The concentrator cartridge is ready for a new experiment. Now the standard solution which contains about the amount of anions we expected is pumped through the column. The volume of standard solution passed through the concentrator cartridge must contain between 1 and 10 μg anions separately. The procedure with standard and sample solution must be repeated three times.

Evaluation

Compare the retention time of chromatographic peak in standard and sample solution, whether the retention time in reference solution and sample is the same as a given value (about 2 to 3%). State which anions are present in the sample.

Calculate the sensitivity factor for anions in reference solution dividing the area by mass of anion (A_i/m_i, where A_i = average area of anion considered, m_i = absolute mass of anion).

The average area of anions present in the sample divided by sensitivity factor gives the absolute amount of anions. Absolute amount of anions divided by volume of eluent pump through the concentrator cartridge gives the concentration of anions.

Determination of Anions in Drinking Water

Theory

All theoretical and practical aspects of ion chromatography can be applied to this determination (see Section 20.1.4). In drinking water, seven anions can be considered: fluoride, chloride, nitrite, bromide, nitrate, phosphate, sulfate, and in some cases iodide. Drinking water is a relatively clean and simple matrix and in most of the cases the sample can be injected directly to the ion exchanger. The only problem is the small solid particles from the pipeline. The solid particles must be removed from the sample by filtering with 0.22- or 0.45-μm pore diameter membrane filter. The second difficulties might arise from humic acid content of drinking water. If it is high, from time to time the column must be regenerated.

Most of the organics can be removed from the drinking water by using a solid-phase extraction cartridge packed by a polar stationary phase.

Apparatus

An ion chromatograph, the same as that used in the determination of anions in high purity water except the injection valve, which is equipped with 20-μl sample loop instead of concentrator cartridge, UV detector connected series to CD. The disposable membrane filter disk diameter is 13 or 25 mm and pore diameter 0.22 or 0.25 μm.

Chemicals

The same eluent and standard solution are used as determination of anions in high purity water: Sep-Pak C-18, solid-phase extraction (SPE) cartridge.

Procedure

Set the wavelength at 215 nm. Take a sample from tape of about 10 cm^3, filter with a membrane filter (pore size 0.22), and force through a Sep-Pak C-18 SPE cartridge. The first part of water is discarded (about 1 cm^3); collect a 2-cm^3 sample into a vial. From this sample inject 20 μl to the column, at least three times.

Calibrate the chromatographic system using a standard solution containing fluoride, chloride, bromide, nitrite, nitrate, orthophosphate, and sulfate 0.5, 1.0, 2.0, 5.0, 10, 20, and 40 μg/cm^3, respectively. Every concentration must be injected twice. The Sep-Pak C-18 or any other SPE cartridge must be activated before use. This can be done by 10 cm^3 acetonitrile followed by 10 cm^3 high purity water forced through the cartridge.

Evaluation

First make the calibration curve both for CD- and UV-detected anions using the chromatographic peak area. By UV detector nitrite, nitrate, and bromide can be detected. Use the calibration curve for evaluating anion contents of drinking water. If some anion concentration is out of concentration range of calibration solutes, dilute the sample by high purity water and measure again.

Determination of Anions in Waste Water

Theory

The principle of ion chromatography is described above. The determination of anions in waste water needs very careful sample preparation. The waste water contains significant quantities of natural, charged organic, and inorganic contaminants. The natural organic can be removed with Sep-Pak cartridge containing a polar stationary phase. Inorganic and organic cations can be removed by treating the sample with anion exchanger. The large pH difference between the sample and eluent can be eliminated by diluting the sample by eluent concentrate or using a special sample preparation column or device (such as Millitrap™ from Millipore). The polyvalent anions must be removed before analysis. It can be done by weak anion exchanger.

Apparatus
See above.

Chemicals
See above. Strong and weak ion exchanger SPE columns (Accell™ Plus QMA, Accell™ Plus CM, Millitrap™ Millipore) are also used.

Procedure
Filter the sample through a 0.22- or 0.45-μm membrane filter; measure the pH of filtered sample. If the 5 < pH < 9, it does not need to be adjusted; if pH < 5 add a small amount of lithium or sodium hydroxide; if pH > 9 dilute the sample by eluent concentrate. For removing the organic use an activated Sep-Pak C-18 SPE cartridge and Accell™ Plus QMA and Accell™ Plus CM. In all cases the first part of sample must be discarded.

For calibration inject a working standard which is described above. If the linearity has been established over a given concentration range, only one calibration standard need be injected.

For sample analysis with the filtered and prepared sample the loop of injector valve must be flushed with at least twice the volume of injected sample. After the injector is switched to the inject position peak retention time(s), area(s), or height(s) are recorded by a data acquisition device or strip chart recorder. The sample must be injected at least three times.

Evaluation
From the retention time the anions present in the sample can be evaluated. From calibration curve or sensitivity factor the amount of anions can be calculated. If the same anions are out of linearity, dilute it and repeat the analysis.

Sample Preparation for Determination of Trace Amounts of Organic Compounds by HPLC

A. Use of SPE for Sample Preparation

Theory
Liquid-liquid extraction is a widely used sample preparation technique, but it has several disadvantages. It is time consuming and needs a high volume of

ultrapure extraction solvent, and selectivity is very limited. For the analysis of organic trace constituents, evaporation of extraction solvent is needed. The higher the concentration factor the higher the concentration of the solvent impurities. Solvent with very limited solubility in water can be used for extraction. For concentration and separation of organic trace pollutants in water, a highly promising approach is to enrich them on the surface of suitable sorbents such as silica, florisil, alumina, alkyl modified silica, porous polymer, modified porous polymer, ion exchangers, and other solids packed in a short plastic, glass, or stainless steel column or cartridge. These small columns can be used as an on-line concentrator, as well, when packed with small-diameter particles used in HPLC or off-line; when the sample preparation is done separately the cartridge is packed with 20- to 40-μm particles and can be operated under gravity flow conditions, or force the liquid through them with vacuum or relatively small pressure drop.

The main functions of SPE cartridge and column are the following:

> Clean-up of sample analyzed
> Trace enrichment
> Storage of extracted solutes

The clean-up function of a sorbent means the partial removal of interfering compounds. The simplest clean-up procedure is when the solutes interested cannot be adsorbed on the sorbent, but the matrix components strongly is adsorbed on it. In this step the concentrated solutes can be separated into different classes by means of a stepwise gradient elution.

The trace enrichment is usually connected to clean-up function as well. Solutes with a strong affinity for sorbent used can be concentrated on the top of cartridge used. The affinity can be controlled by the surface chemistry of stationary phase. The surface chemistry of sorbent used depends on solutes analyzed. One of the main classes of these sorbents is based on modified silica gel. Much modification and immobilization is known to prepare a sorbent which has a good affinity to solute concentrated. Nearly all type of sorbent can be used for sample preparation which are used in HPLC. Horváth et al.[38] gives a good overview of the stationary phases and immobilization of compounds for preparing of selective adsorbents. Alkyl-modified silica is widely used for enrichment of polynuclear aromatic hydrocarbons (PAH).[107-115] Various adsorbents tested for extraction of PAH showed different recoveries. Some authors stated there are great differences from lot to lot as well.[16] Recovery of different PAH varied from 15.6 to 89.6% for naphthalene and fluoranthene.[109] Solubility of PAH in water is very low, about 0.001 μg/dm[116] for benzo(a)pyrene. In surface or drinking water the PAH are present in higher amounts. They adsorbed onto solid particles and solubilized by surfactant. Both the water and solid particles must be analyzed to get a clear picture about PAH concentration in surface or other types of water.

To get high selectivity and sensitivity HPLC equipped by fluorescence detector must be used. The detection limit is about 0.1 ng/cm^3 for standard solution.

Apparatus

A gradient elution liquid chromatograph equipped with fluorescence detector

    Filtering device
    Graduated cylinder, 500 cm^3
    Volumetric flask 10 and 25 cm^3
    Vacuum box
    Rotary evaporator

Chemicals

    Acetonitrile, chromatographic grade
    High purity water
    PAH standards
    SPE cartridge C-18, ODS, or other octadecyl modified silica

B. Procedure for Determination of PAH in Drinking Water, Surface Water, and Waste Water

Sample Preparation

Take the SPE cartridge and wash with methanol of 10 cm^3 and 5 cm^3 of 10 v/v methanol/water mixture. Take a 500-cm^3 sample from the drinking water and force through it by vacuum box at least five samples of about 8 cm^3/min. Wash the cartridge with water and elute the PAH with 2 cm^3 acetonitrile. The organic solution can be injected directly onto the chromatographic column or can be concentrated with nitrogen if necessary.

Sample Preparation for Determination of PAH in Surface and Waste Water

For determination of PAH in surface and waste water, the sample preparation must be divided into two parts. The samples must be filtered as soon as possible after the sampling. The filtered water sample treatment must be done as described for drinking water. The PAH must be dissolved from the solid particles by acetonitrile (10 to 50 cm^3). Cut the filter paper as small as possible and place it into a beaker and pour acetonitrile to just cover it. Put it in an ultrasonic bath for 10 min. Filter the sample, and dissolution of PAH must be repeated. Join the acetonitrile solution and evaporate in rotary evaporator in vacuum about 0.5 cm^3. This solution can be injected onto the chromatographic column.

Chromatographic Conditions

The chromatographic conditions for separation of the PAH have been investigated in Section 20.3. According to our experimental work, one of the best columns for this purpose is a Bakerbond PAH column 16-plus, 250×3.0 mm and gradient elution must be done with the following parameters:

Time (min)	vol.% of acetonitrile in the eluent
0–5	50 (eluent A)
3–35	50–100 (linear)
35–45	100 (eluent B)

The fluorescence detector parameters are $\lambda_{EX} = 240$ nm, $\lambda_{EM} = 400$ nm.

In the following part of this description we will give complete validated methods for determination of pollutants in water. This procedure is one part of quality assurance in environmental chemistry.

Sampling

The samples must be analyzed after the sampling. If the samples have to be stored, add 2 cm^3 of concentrated sulfuric acid to the samples and 1 g of copper sulfate, and store them in a dark place at about 0°C. A detail description of sampling sites and procedures must be given including the methodology, labeling, container preparation, storage, and pretreatment of procedures. An acceptable sampling should include the following:

1. Proper statistical design, which takes into account the goals of studies and its uncertainties and certainties. In our case the main task is determination of benzo(a)pyrene. According to the old World Health Organization (WHO) rule at 10 ng/dm^3, the accepted new one is about 700 ng/dm^3.

2. Instruction for sample collection, labeling, preservation, and transport to the analytical facility.

3. Training of personnel in the sampling techniques and procedures specified. Sampling for determination of PAH [benzo(a)pyrene] in drinking water seems to be relatively easy. However, none of the rules can be canceled. During storage do not alter the composition of the substance to be analyzed. It means 10 and 700 ng of benzo(a)pyrene must be taken to a blank water, and to distillate water. Time to time analysis of the spiked waters must be taken.

Sampling for determination of PAH in surface water is much more complicated. The samples have to be taken at the surface or from under surface. The samples have to be filtered and water and solids filtered must be stored until analysis. During the transportation the solved and adsorbed amount of PAH may be changed and so on.

Sampling for determination of PAH in waste water is even more complicated. Which time and place and how must be done the sampling? Preserving of the samples must be done on the spot; otherwise, some biological activity is always present in the waste water and may alter the composition.

The more difficult the sampling, the higher the number of the samples have to be taken to control and characterize the sample and enhance the reliability of the final results. It is out of the scope of this description to give detailed information about it. The interested reader is referred to the

"Guidelines for Data Acquisition and Data Quality Evaluation in Environmental Chemistry" which was published in *Analytical Chemistry*.[137]

Sample Pretreatment

Sample pretreatment is involved in physical operations. In our special case the filtering must be documented, giving detailed information of the sample history, so that another worker can exactly duplicate the treatment used.

Calibration and Standardization

Calibration in determination of PAH in different water means the determination of the response function S = g(C), where S is the measured signal which is a function of "g" of given analyte concentration. For this purpose at least five different concentrations of calibration standards should be measured in triplicate. The concentration chosen should bracket the accepted concentration of the solutes in the field sample. Response factors and calibration curves must be determined varied amount of solutes in a suitable solvent or much more valuable in a sample matrix (blank). In our cases the calibration for all PAH interested means the calibration curve and response function can be the determination.

Method Testing and Monitoring

Precise measurement depends largely on the proper use of good laboratory practice, methodology, and instrumentation used. Accuracy will be determined by the purity of the standard used (purity) and stability (change during time by decomposition or adsorption on the wall and so on). The purity of reagent must be certified. Usually the ruggedness test is to be controlled from time to time. The stock solutions must be controlled occasionally with a freshly prepared standard.

Any proven method must be retested during the measuring process by periodic analysis of the blanks (any blanks which may cause artifact results or interference), standard (concentration may change through decomposition of substances, evaporation of solvent used), and "spiked" samples (to monitor of recovery and specificity of method used). Selectivity of method will prove a guideline for evaluating testing results (a knowledge about the decomposition products of standard, matrix components cause enormous results and so on). If any critical condition is monitored, steps should be taken to avoid anomalous results.

In determination of PAH in drinking water, surface water, and waste water, after the sampling the water samples must be divided into two parts. One part serves as a control or testing sample; the other samples are used for the analysis. To the control sample add benzo(a)pyrene solved in ethanol about (0.1 mg/cm³) 0.5 cm³ to 1 dm³ sample. To monitor a proven method reliability, these samples are used.

Definition of the Data Set

Trace analytical measurement (determination of PAH in water) requires that data obtained from the following categories:

1. Calibration standards (origin, certified content storage conditions, and evaluation of purity).

2. Field blanks for determination of PAH in water are very difficult to find; appopriate field blanks contain every matrix component except for PAH.

3. Spiked field blanks (as described above).

4. Spiked laboratory blanks; add distillate water the same amount of benzo(a)pyrene and checked it time to time.

5. Working standards; used for determination of response factors, if they are stable and recovery of PAH is reproducible, calibration will not do it as frequently as otherwise necessary.

6. Field samples; samples must be analyzed.

Determination of PAH in water needs multiple manipulations and therefore this will increase measurement variability; regular testing with the spiked field samples must be conducted.

Performance and Data Testing

Electronic data handling, data reduction, and data storage greatly facilitates data handling and helps to control the data evaluation, but it must be tested periodically. Recalculation of known data must be done to prove the reliability of data handling. Data calculated one day must be repeated the second day and compared.

Measurement Variability

In every analytical measurement two types of errors can influence the final results: systematic and random errors. If a measurement carried out using a reliable working standard but excessive measurement variability is found, uncontrolled systematic errors must be found. For example, the packing material of a solid-phase extractor cartridge will not be suited to sample preparation if its surface chemistry is not controlled by the manufacturer. In a validated method all practical steps have been controlled according to the systematic errors (suppress, eliminate, or compensate). Random errors arise due to weighing uncertainties, aliquoting variability, or sample inhomogeneity; all fluctuations that are not considered as a systematic error will determine the experimental precision.

The source of systematic errors in determination of PAH can be considered: sampling, storage, sample preparation, and HPLC determination method used.

The limit of detection, limit of quantitation, recovery, and other terms used for validation can be found in Sections 3.3 and 3.6.

Qualitative Confirmation of Validated Measurement

This is based on an analytical principle of different analytical conditions different from those used in the initial method. PAH determination with HPLC-fluorescence must be confirmed by GC-MS with selected ion monitoring.

Verification

The reliability and acceptability of analytical information can be judged on a well-defined protocol. In the protocols all information which influences the reliability and acceptability of results has to be documented. Such protocols must describe the documentation requirements of the study, including sampling procedures, measurements, validation, and confirmation. The results must be reviewed critically. If any doubt about the reliability of results arises additional confirmatory tests must be done. If a result differs highly from results of the large population, determination must be duplicated and applied with an independent method. Agreement of the three results indicates reliability of the measured value.

Documentation

Any new methology worked out must be described in detail. The use of existing methology can be cited by reference to published literature, but any modification has been fully tested and reported.

Raw data must be given such as sample number, initial amount, extraction volumes, final weight, volume injected onto chromatographic column, instrument response, concentration of samples and standards, and so on.

Evaluation

Present the following parameters: limit of detection, limit of quantitation, linearity, precision, recovery, day-to-day reproducibility; and make a complete documentation from measurement including the concentration of PAH in drinking, surface, and waste water.

Determination of Phenolic Compounds in Drinking Water, Surface Water, and Waste Water

Theory

There is an increasing need for trace-level determination of phenol and substituted forms in aqueous environmental samples. Phenol is a highly polar compound and the sample preparation may not be done by using the same approach which was done at the determination of PAH. The chlorinated phenols produced during the water treatment are more hydrophobic and can use the same approach as was used for PAH. In surface water and more likely in waste water, there are many interfering unknown matrix compounds that influence the sample preparation. For these matrices, a preseparation step must be taken.

Distillation of the phenolic compounds is a good preseparation technique because most of the matrix compounds are not volatile and most of the anions and cations will remain in the distillation flask. Combined electrochemical and multiwavelength detection offer a good chance for identification and validation of method used. An alternative on-line sample treatment can be applied when the analyte concentration is low; it can be done from drinking water and after the distillation of the surface and waste water.

It has been demonstrated that carbon-based materials and copolymer-based sorbents have an enhanced affinity for medium-polarity solutes such as mono- and dichlorophenols. Graphitized carbon black (GCB) acts as a strong reversed-phase sorbent; HPLC experimental work has demonstrated that the affinity of GCB to phenolic compounds is nearly two orders higher than alkyl-modified silica gel (C-18) and between three and four times higher than copolymer-based sorbents.

Apparatus

> Isocratic or gradient HPLC system equipped with electrochemi-
> cal and multiwavelength detector
> Distillation equipment
> Graduated cylinder, 500 cm^3
> Ultrasonic bath
> Weighing machine
> Glass bottles, 1000 cm^3
> Volumetric flasks, 25 cm^3 and 100 cm^3
> Reversed-phase column, 5 μm (preferred Nova-PAK C-18 car-
> tridge, 100 × 3.9 mm, d$_p$ = 4 μm)

Chemicals

> Phenol, chlorophenols, methyl phenol standards
> Acetonitrile
> Sodium perchlorate
> Trisodium citrate
> Acetic acid
> PRP-1 or other copolymer-based sample preparation cartridges
> Hypercard porous graphitic carbon
> Water
> RP-18 cartridge for sample preparation
> Methanol

Procedure

Sample Preparation for Determination of Phenolic Compounds in Drinking Water. Sample taken from running water into 1000-cm^3 glass bottle; if it is not used immediately after sampling, 2 cm^3 of concentrated sulfuric acid and 1 g copper sulfate must be added. The preserved samples can be stored in a dark place and about 0°C. The sample of about 100 cm³ (but it depends on the concentration of phenolic compounds interested) must be forced through a coupled sample preparation cartridge. First is a packed C-18 (RP-18) modified silica gel; the second is RPR-1 or GPC material. Both columns must be washed with methanol before use. The first column will adsorb the hydrophobic phenols. The second one will retain the polaric and moderately polaric phenols. Wash the column with high purity water. Elute the phenolic compounds with about 2 cm³ of methanol containing 1% acetic acid. This solvent will be used for the analysis.

The same sample preparation must be done with the spiked sample in the range of 10 to 100 µg/dm³ for each of the solutes measured. From surface water and waste water a preseparation step must be taken. Pour 100 cm³ of a sample in a round-bottom flask connected to a still head and condenser (Graham-type). Collect the first 90 cm³ in a volumetric flask, stop the distillation, and after cooling down add 10 cm³ of water and start the distillation again until 100 cm³ distillate sample is collected. This distillate can be injected directly onto the chromatographic column or concentration must be followed as described above.

Chromatographic Conditions.   The separation of phenolic compounds can be done either with isocratic or gradient elution technique. If phenol, mono-, and disubstituted derivatives must be determined, eluent composition is about 40/60 v/v acetonitrile/buffer, buffer contains 0.2 M sodium perchlorate, 0.005 M trisodium citrate, and 1% acetic acid. If a broad range of phenolic compounds (hydrophobic phenols such as pentachlorophenol) must be determined, the application of a gradient system may be advisable. The eluent A is the same which must be used for isocratic separation, eluent B is 80/20 v/v acetonitrile/buffer sodium; perchlorate concentration is 0.01 M. The preferred gradient shape is linear; gradient time is 35 min at flow rate of 1.5 cm³/min. For the separation of phenolic compounds a reversed-phase column can be used; the preferred one is a Nova-PAK C-18 cartridge (100 × 3.9 mm, d = 4 µm). Electrochemical detection can be done at 1100 mV (vs. Ag/AgCl); the wavelength range for multiwavelength detector is between 200 and 320 nm.

Method Validation.   Follow the instruction given on determination of PAH in Sections 3.3 and 3.6.

Evaluation.   Follow the instruction given on Sections 3.3 and 3.6 on determination of PAH, and give the concentration of phenolic compounds in drinking, surface, and waste water.

20.4.3. Analytical Application of HPLC in Clinical Chemistry

Theory

Electrochemical detection (ED) is widely used in clinical chemistry. About 60% of the application of ED has been done in the field of bioanalysis. The main advantages of using ED are the selectivity and sensitivity. The principle of ED used in biomedical analysis is a transfer of charge between substances in a column effluent and a working electrode. The amperometric detector works at a constant potential against a reference electrode. The current through the working electrode is the signal which is measured. The term amperometric detection is connected to one of the mostly used ED, when only a fraction of the solutes is oxidized or reduced. The coulometric detection is considered when the electrochemical conversion is complete or nearly complete. About three fourths of amperometric detectors used in practical application work in the oxidative mode. The dissolution of oxygen and limited range of electrode materials used in negative potential makes this type of ED less popular.

In amperometric ED several cell designs have been tested, but practically only the thin-layer and wall-jet cell with a three-electrode configuration are used.[138-148] In amperometric detection the electrochemical reaction takes place on the surface of working electrode; the signal does not depend on the volume of the cell. In thin-layer design where a thin spacer is placed between the two rigid blocks, the volume of the cell can be reduced by decreasing the thickness of the spacer. The volume of the cell is usually between 1 and 5 μl, but thin-layer cells with volumes of 10 to 100 nl have been also reported. The body of the cell is made of an insulating material such as PTFE, PEEK, or other plastic. The electrode material used in ED is normally carbon in one of its several forms. At present the glassy carbon is the most commonly used electrode material, which is mechanically strong enough, not attacked by organic solvent, and easy to clean. The reference electrode is usually placed close to the working electrode (Ag/AgCl or calomel). The auxiliary electrode serves to carry the current and is made of platinum or sometimes stainless steel. The thin-layer cells or workings electrodes can be coupled. In a parallel dual electrode configuration two working electrodes work independently of each other at different potentials. If one electrode is at a negative and one at a positive potential, oxidizable and reducible solutes can be measured in one chromatographic run. To improve the detection selectivity, one electrode is used at lower potential and the other at higher potential. At lower potential only the most easily oxidized solutes can be detected. With two parallel electrodes a peak purity assessment can be done, because the ratio of current of a solute at two different potentials is a constant value and independent from the history of electrodes.

In series dual configuration with the first working electrode the solute interested can be reduced, while the second is applied in the oxidative mode for measuring the reduced solute.

In wall-jet design the effluent and solutes can reach the surface of the working electrode as a free stream. The jet (end of the capillary tube connected to the column outlet) diameter must be less than the diameter of the working electrode; the gap between the end of the capillary tube and working electrode must be enough that the tip does not interfere with the reflected flow from the electrode surface. The effective cell volume is hydrodynamically limited to a thin layer over the surface of the working electrode.

The amperometric detector can be worked in pulse mode as well. In pulsed amperometric detection (PAD) mode the irreversibly adsorbed compounds can be removed and compound otherwise electrochemically inactive can be detected on an activated electrode surface. With a modern-day amperometric detector the hydrodynamic voltammogram can be measured used for a selective and sensitive detection in biomedical application.

The electrochemical (amperometric) detection mode is used for determination of easily oxidizable solutes. The potential window is usually limited to 0.1 to 1.5 V (vs. Ag/AgCl). The reduction of the hydrogen ion will be the lower limit and oxidation of eluent and its impurities the higher limit of the potential window. At the potential window given above catechols, phenols, thiophenols, aromatic amines, α-oxo-acids, α,β-unsaturated carbonyls, tertiary alkyl amines, thyohydrations, phenothiazines, and carbohydrates can be detected.

In biomedical applications the advantage of ED is the improved selectivity and sensitivity over the generally used UV-VIS detection. For determination of catecholamines in urine or other tissues only this detector can offer high sensitivity and selectivity. The limit of detection for catecholamines in urine is about 1 μg/dm^3 or even lower. With the fluorescence detector the same detection limit might be reached, but not all catecholamines interested in clinical chemistry can be detected. To make a fluorescent-active derivative is time-consuming and laborious.

In the literature several hundred publications deal with tyrosine metabolites (catecholamines, o-methyl derivatives of catecholamines, acid and neutral metabolites of tyrosine), tryptophan and its metabolites, aromatic amines (benzidines, aniline-derived carbamates), ascorbic acid, uric acids, and different drugs and their metabolites. Some of them are the following: tricyclic antidepressant, penicillamine, amoxicillin tocopherols, morphine, benzodiazepines and their metabolites, etc. Sulfur-containing compounds and some inorganic ions can be detected by ED.

In HPLC most of the applications have been carried out by UV-VIS detector and ED only used in small portion, but in bioanalytical application use of this detection method offers advantages over the preferred UV-VIS one.

Homovanillic acid (HVA) and vanillylmandelic acid (VMA) are commonly measured in urine for the differential diagnosis of neuroblastoma, pheochromocytoma, and related tumors. HVA has been used to monitor chronic lead exposure and response to medication during the treatment of Parkinson's disease. Usually abnormalities of biogenic amine metabolism are implicated in various pathological states including psychiatric and neurological disorders.

The development of HPLC with ED facilitated the establishment of a highly sensitive and selective assay procedure for these compounds. Determination of VMA and HVA needs a careful sample preparation. Using an ion exchanger for sample preparation will simplify of analysis of urinary VMA and HVA.

Determination of VMA and HVA in Urine

Apparatus

Isocratic chromatographic system equipped with pulsation dampener

Amperometric or coulometric electrochemical detector and variable-wavelength UV detector

pH meter with glass electrode

Ultrasonic bath

Weighing machine

Centrifuge

Plastic container, 10 cm^3

Graduated cylinder

Membrane filter device with a membrane filter
$d_p = 0.45$ μm

Spherisorb ODS or other reversed-phase column with similar retention and selectivity characteristic, quartz cartridge (to increase the lifetime of column used)

Chemicals

VMA and HVA standards, ethanol, 2-propanol HPLC reagent grade

0.05 M phosphate buffer, pH = 3.0; preferred from 1 M potassium dihydrogen phosphate and 2 M phosphoric acid

6 M hydrochloric acid

Ion exchanger, preferred AG1-X2 (BioRad 140-1241)

2 mM i-VMA

3 mM i-VA

0.05 M citric acid

Sodium acetate

1.5 mM octane sulfonic acid sodium salt

0.15 mM Na$_2$EDTA

Procedure

Sample Preparation.   The first step of determination of VMA and HVA is the correct urine collection and clean-up procedure. The daily urine must be collected in plastic containers with 10 cm^3 of 6 M hydrochloride acid as a preservative. For longer periods, the samples must be stored at –4 or –20°C. For a shorter period (less than 1 week) they can be stored between 0 and 5°C. To remove the solid particles from the collected sample, samples must be centrifuged at least at 2000 g. The procedure is the same if we want to determine the VMA and HVA content in a given sample.

From the collected sample 5 cm^3 is taken into a plastic container and 100 mm^3 2 mM i-MVA and 3 mM i-VA and 10 cm^3 0.08 M sodium acetate solution (pH = 6.1) must be added. The pH must be adjusted to 6.1 by 0.5 M sodium hydroxide.

The exchanger must be conditioned before use by 20 cm³ sodium acetate, and the sample is allowed to drain. To remove weakly sorbed matrix components the column must be washed with 20 cm³ sodium acetate buffer.

The HVA and i-VA must be eluted with mixture of 4.5 cm³ ethanol. The VMA and i-VMA must be eluted with 10 cm³ 0.5 M phosphoric acid. The eluates can be stored in the refrigerator for 24 hours. For measurement the eluted samples must be diluted; the usual dilution rate is five.

Preparing of Eluent. For the determination of HVA a 0.05 M citric acid sodium acetate buffer (pH = 4.3) containing 1.5 mM octane sulfonic acid sodium salt, 0.15 mM Na$_2$EDTA, and 3 v/v acetonitrile must be prepared. An alternative eluent used by BioRad Laboratories in Richmond is 0.05 M phosphate buffer pH 2.30, 100 cm³ ethanol, and 20 cm³ 2-propanol in 1000 cm³ buffer solution (eluent A).

For the determination of VMA a 0.05 M citric acid-sodium acetate buffer (pH = 3) containing 0.15 mM Na$_2$EDTA must be prepared from 2 M phosphoric acid and 1M potassium dihydrogen phosphate and 16 cm³ ethanol in 1000 cm³ buffer solution (eluent B).

HPLC Operating Instructions. For determination of HVA with eluent containing ion-pairing reagent a careful equilibration can be done. Using this mobile phase not only the HVA but other biogenic amines can be determined in one run. For HVA and VMA determination the measuring potential is 800 mV (Ag/AgCl) but it might be different if somebody uses as coulometric cell an ESA cell (Model 5100 A Coulochem, Environmental Sciences Associates, Bedford, MA). Through the hydrodynamic voltammogram the measuring potential must always be checked. A multiwavelength UV detector is connected after the ED, set at 228 nm or 280 nm. If the background current during the run will increase highly, disconnect the column from the ED, and wash 25 cm³ water and 25 cm³ 50/50 water/ethanol mixture and again with water. Running the HPLC equipment overnight with the flow rate at 0.1 cm³/min is advisable.

Validation of Method

1. Linearity — Make a standard solution containing 1 to 100 nmol of VMA, i-VMA, HVA, i-VA, and inject into the chromatographic column at least twice.

2. Limit of detection (LOD) — From the standard solution to the certified urine sample 50 nM compound interested in and after sample preparation and dilute until the signal is about three times higher than the noise level.

3. Limit of quantitation — The procedure is the same as written above; the dilution must be done until the standard deviation of measurement is about 5%.

4. Recovery — HVA, i-VA, VMA, and i-VMA were added to a certified urine sample in different concentrations from 10 to 100 nM. The absolute recovery of HVA and VMA must be determined by assigning these samples as described and comparing the peak area of VMA and HVA obtained from injection of a standard solution of VMA and HVA in eluent. For each concentration at least three different independent measurements must be done. To get data for standard deviation of determination, take five urine samples (5 cm^3) and add 50 nmol of solutes to them and inject twice onto the chromatographic column.

5. Precision — A certified urine sample was taken and compounds were added, at concentrations from 10 to 60 nM and after sample preparation injected onto the column one day at least seven times and from the same solution repeat this procedure during a week.

6. Selectivity — To a certified urine sample take 50 nM compounds and after sample preparation measure the peak area at two different potentials: 500 and 800 mV and at 228 and 280 nm.

Evaluation. Calculate LOD and limit of quantitation and its standard deviation, plot peak area and peak height against concentration compound interested, calculate recovery and its standard deviation, calculate reproducibility within day to day, and calculate the concentration of VMA and HVA in a sample taken from a healthy adult and an ill patient.

Determination of other catecholamines can be done with a similar approach; separation of some of them is described in Section 20.1.3.

Determination of Different Drugs in Serum

Theory

The drug monitoring offers an important guide to clinical therapy. The relationship between drug concentration and therapeutic and/or side effects is complicated by the presence of metabolites. Monitoring of metabolites will be an essential part of pharmacokinetic studies. In a biological fluid several compounds are found. To validate a method for determination of drugs and their metabolites in biological fluids based on detection methods which give information for peak purity and peak homogeneity. The multichannel detection of these widely used methods for evaluation of peak homogeneity and purity with a conventional variable wavelength detector a multiple injections must be made, each at a different wavelength chosen for the various solutes in a solution. With a multichannel detector so-called total information can be obtained. The information matrix absorbance (A), wavelength (λ), and time (t) can be presented as chromatograms at different wavelengths as spectrum at different times and from the spectral data the purity of chromatographic peaks can be obtained.

The photodiode array (PDA) detector has two main parts. One is the optical unit, which consists of a radiation source, sample cell, diffracting grating polychromator, and a multichannel detector which is usually a linear PDA. The multiwavelength detectors generate a huge amount of data (A, λ, t). To collect data and get analytical information and make a post-run evaluation sophisticated software can be used. The computer (PC) and special software is the second part of PDA. Most commercially available PDA detectors utilize reverse optics.[149-153] All the light from the radiation source is focused through the flow cell and the transmitted light is dispersed by the polychromator (holographic grating or prism) to the PDA.

Several methods can be used for determination of peak purity and homogeneity. The chromatogram and spectral overlays at different characteristic wavelength and at different time of eluting chromatographic peak are the simplest way to validate peak purity and homogeneity. The data matrix can offer the following possibilities to evaluate the peak purity and homogeneity in bioanalytical application where exogenous and endogenous compounds are present:[154]

> Absorbance ratio (ratio plot) method
> Spectral suppression of one or more known components
> Second or higher derivate of absorbance in time domain
> Normalization of spectra recorded on the leading and tailing
>   edges and at the apex of a chromatographic peak
> Spectral deconvolution as a function of elution time
> Principal component and factor analysis to assess the most
>   probable number of overlapping peaks in a cluster

With a spectral archive retrieval system solute recognition can be achieved. This stored spectrum for standard material can be compared with solute in a mixture; chromatographic runs were done at the same mobile phase composition.

To prove the utility of PDA detector over single channel detector for determination of antidepressants we will use a PDA and single-channel detector in a series. The first step of analysis is the collection and storage of samples. The antidepressants are hydrophobic amines. They can adsorb onto the surface of any hydrophobic container or on an acidic surface. Plasma can be collected in heparinized glass or plastic tubes. Some plastic tubes contain plasticizer tris (buthoxethyl) phosphate, which leached out into the blood samples and was found to displace the basic drugs such as imipramine and its metabolites from their binding to α_1-acid glucoprotein.

The drugs were sorbed by red blood cells, causing the measured concentrations to be erroneously low in the separated plasma or serum. At low concentration levels, the effects of collection vessels and their contaminants in the blood sampling may influence drug analysis and need to be evaluated. Recently separator gel is being used in a collection tube. The concentration of imipramine and its metabolites were found about 1/5 compared to separator column not exposed to the separator gel.[155] The decrease in drug concentration separator gels is likely caused by binding of drug to the gel. In addition to the adsorption to the serum separator gels, the drug can be adsorbed onto glassware during storage and sample preparation. In some cases drug adsorption to chromatograph was reported as well. Deactivation of glassware surface can be done chemically to collect the serum or plasma. Plasma and serum are assumed identical for determination of drug concentration. However, it might be true or not; the process by which the plasma and serum are prepared may affect the measured drug concentrations. The measured concentrations of tricycles anti-depressants are inconsistent when concentrations determined in plasma and serum are compared. The mean concentrations of imipramine in heparinized plasma samples were 32%[156] and 5%[157] higher than in serum, while the concen-tration of desipramine, amitriptylene, and nortriptylene were about the same in both serum and plasma samples. In literature some reports are confusing since there is no real indication about which biofluids were analyzed. At the sample treatment a clear identification must be made about the sample collection procedure.

The effect of anticoagulants on drug concentration measured has not been investigated in detail. Such effect of endogenous trace elements has been reported already.[158] For example, heparin is known to activate lipoprotein lipase, citrate anticoagulant induce leakage of intracellular water from red blood cells, and cause a significant dilution, lowering the drug concentration measured. In addition to this during storage enzymatic degradation might affect the drug and its metabolite concentration. In a pharmacokinetic study, effects of anticoagulant used must be investigated.

Biological samples are multicomponent mixtures, in which drugs are usually present as the minor component. Isolation and concentration of the analytes need a preconcentration procedure. Solvent extraction, solid-phase extraction, and on-line solid-phase extraction using a column switching technique and protein precipitation can be used for isolation and enrichment of drugs in biological samples. Solvent extraction has some disadvantages compared to solid-phase extraction, but in clinical chemistry huge amounts of hydrophobic drugs to be analyzed might be favorable. Direct injection of biological samples for HPLC analysis can be done for specialized columns. Pinkerton[159] has described a novel approach for pretreatment of plasma samples. The stationary phase used for direct injection of plasma contain a polar external surface and the pores are hydrophobic; the large plasma protein molecules cannot penetrate into the small pores (size exclusion effect) and elute from the column in a short time the travel through is the interstitial volume. The analytes can penetrate into the hydrophobic pores and adsorb on it. The increasing coverage of the hydrophobic surface by proteins is avoided and no limitation the number of samples are injected into the column compared to conventional reversed-phase column where gradual adsorption of protein molecules makes them ineffective. The direct injection of plasma without protein precipitation can be done for column or be done for column coupled with a guard column or a wide pore column as well. Direct serum or plasma injection can be done if the concentration of analytes is high enough, compared to the limit of detection, and eluent contains more than 20 to 40 v/v water or buffer. At high organic concentration the protein molecules may participate in the column when serum and plasma will be mixed at the inlet of column. Direct injection of whole blood is another approach. The measured drug concentration may be higher compared to plasma and the serum one.

In most pharmacokinetic studies the collection the biological samples and the determination of drug concentration is seldom done immediately. The serum and plasma samples must be stored until the analysis is performed. Long-term storage enhances the possibility of degradation of the drugs and their metabolites. In addition to this drug metabolism may be generated during the preparation of the samples. Analyzing plasma and serum samples at peak purity and homogeneity is a critical question. One must be sure no other degradation product or artifacts are produced during the collection, processing, and storage. The drug analyzed must not interfere with any endogene or other compounds. The statement above is very important because many drugs and their metabolites in low concentration are relatively unstable. There are several factors that may cause degradation. Blood specimens collected from man and animals must be centrifuged to prepare plasma or serum samples. Time and temperature before centrifugation may cause variation in drug concentration measured. To get a correct result for some drug concentration centrifugation of the blood samples must be done immediately after collection. The so-called "blood storage effect" means the redistribution of drug by diffusion across the erythrocyte membrane. For transporting serum or plasma they must be frozen in liquid nitrogen. Drug and metabolites may be degraded chemically, photochemically, or enzymatically. The photochemical degradation can be avoided by protecting the samples from direct UV or sunlight. The enzymatic degradation must be determined for drug and its metabolites. In whole blood samples some metabolites of drugs may convert to parent drugs or other metabolites and thus may result in an enormous concentration of measured parent drugs or the metabolites. Chlorpromazine, which is a trycyclic antidepressant, will partially oxidize in basic condition. It has been reported that chlorpromazine N-oxide was reduced to chlorpromazine in plasma with sodium hydroxide, but this conversion was not observed in plasma with sodium carbonate. Some chemicals used in sample collection may cause redistribution of drugs between plasma and red blood cells, because they can displace drugs from their binding to proteins or other macromolecules.

Sample degradation during its treatment can be controlled by using an internal standard chemically similar to the drug and its metabolites measured.

Validation of drug determination in plasma or serum uses the following procedures and determination prior to analysis:

Calibration and linearity
Selectivity, specificity, and interference
Sensitivity, limit of detection, and limit of quantitation
Precision and accuracy
Extraction recovery
Quality control and ruggedness of analytical methods
Comparison of performance of analytical methods and
   laboratories
Analytical data system and management

It has been demonstrated that so many factors have effects on a drug and its metabolites concentration, in an analytical procedure every step must be well documented and controlled. If it is not taken the results do not reflect the real situation and the interpretation will be useless.

In our determination of trycyclic antidepressants in drugs, plasma, and serum, all the above-mentioned parameters may influence the measured concentration. The validated chromatographic method is only one parameter to get correct results; sample preparation is the other and sample handling is the next one. If one is incorrect the analytical results will be useless for practice.

Apparatus

Isocratic HPLC equipped by UV-VIS and PDA detectors
Graduated cylinders
Weighing machine
Centrifuge
Nitrogen bomb
Glassware for collecting and storage blood samples, serum or
   plasma (10, 20 cm^3)
Reversed-phase column for basic drugs (C-18, ODS)
Vortex mixer
Pasteur pipette

Chemicals

Imipramine
Desipramine and other imipramine derivatives which are
   commercially available
Chlorimipramine as an internal standard
n-Heptane
Isoamyl alcohol
Sodium hydroxide
0.05 *M* sulfuric acid
High purity water

Buffer (0.2 *M* boric acid, 0.2 *M* potassium chloride, 0.18 *M* NaOH)

Isoamyl alcohol, methanol, acetonitrile, phosphoric acid, trimethylamine, solid CO_2

Procedure

Sample Collection.   In a glass tube containing citrate or heparin the whole blood must be collected and centrifuged immediately after collection and stored at −20°C in a refrigerator until analysis.

Sample Preparation.   1 cm³ serum or plasma in 10 cm³ glass tube must be alkalinized by 1 cm³ buffer pH = 10; add the 100-µl internal standard (1.00 µg/cm³). Extract it with 5 cm³ n-heptane-isoamyl alcohol; to improve the extraction use a Vortex mixer to shake the sample for 5 min. Centrifuge at 1000 to 2000 g for 10 min. The organic phase must be transferred with a Pasteur pipette into another tube containing 1 cm³ 0.05 *M* sulfuric acid. The tube must be shaken mechanically with Vortex mixer about 5 min and centrifuged about 3 min at 1000 to 2000 g. The tube must be dipped into a mixture of methanol and solid CO_2 to free the aqueous phase and the organic phase must be discarded. The aqueous phase must be left to thaw at room temperature; add 0.5 cm³ 1 *M* NaOH and 5 cm³ n-heptane-isoamyl-alcohol. The tube must be shaken mechanically about 5 min and the upper organic layer must be transferred into a conical glass tube which contains 100 µl acetonitrile and evaporated to nearly dryness under nitrogen at 40°C. Add 200 µl mobile phase to the residue and measure of volume with a microsyringe.

HPLC Conditions.   The mobile phase composition is as follows: 1% triethylamine (pH = 3.0 adjusted with phosphoric acid) and acetonitrile 35/65 v/v. The eluent composition depends on column used. Before starting the analysis using standard solution the retention must be adjusted by changing the ratio of buffer and acetonitrile. The preferred capacity factor is between 2 and 5. The single channel and/or multichannel detector (PDA) must be connected in series.

Evaluation

The wavelength for single channel detector is 254 nm; the wavelength range for PDA is 210 to 350 nm.

Calibration and Linearity.   Prepare a stock solution of 100 ng/cm³ from each analyte and internal standard. Dilute the stock solution to get 1, 20, 40, and 60 ng/cm³ standard solutions. Inject at least twice from each solution.

Selectivity, Specificity, and Interference.   Take a drug-free sample and do the sample preparation as described above and inject the sample into the chromatographic column; immediately after this inject a standard solution which contains both parent compound and its metabolites. Compare the two chromatograms; if peak or peaks can be seen at the retention time of reference solutes, adjust the mobile phase until there is no interference with solutes in the sample. Then inject the sample and use a multiwavelength detector. Make spectral overlays, isoabsorbance plot, absorbance ratio, and any other peak purity parameters which can be used for evaluation for the peak homogeneity. If the peak purity test fails adjust the mobile phase composition until a good peak homogeneity parameter can be obtained.

Sensitivity, Limit of Detection, and Limit of Quantitation. Dilute the standard solution until the signal/noise ratio is about three. At this solute concentration inject the samples at least three times. Inject a sample about five times higher than detection limit and repeat it at least five times. Add the solutes to drug-free serum samples and after sample preparation repeat the procedure described above.

Precision and Accuracy. Add the solutes to the drug-free serum at a concentration above the limit of quantitation and inject it into the chromatographic column and repeat it at least five times a day, day after day for at least five different days.

Extraction Recovery. Add the known concentration of solutes analyzed and internal standard to the drug-free samples above the limit of quantitation. At least five different concentrations and five different injections at each concentration must be done.

Determination of Relative Sensitivity Factor. Take a given amount of internal standard and different concentration of solute analyzed to a drug-free serum or plasma sample and inject it at least three times at each concentration.

Data Must be Given. Calculate the limit of detection, limit of quantitation, day in and day to day reproducibility, relative sensitivity factor, extraction recovery, and its standard deviation at different serum or plasma concentrations. Give the purity parameter, calibration curve, and linearity range for solutes analyzed.

Determination and Sample Preparation of Retinoids in Serum or Plasma

Theory

The term "retinoid" includes both the naturally occurring compounds with vitamin A activity and synthetic analogs of retinol. Vitamin A (retinol) is essential for normal growth, vision, reproduction, and so on. The retinoids are very sensitive to light and oxidation. Therefore, special care has to be taken from the beginning of sample collection until the final step. A general description can be found in References.[160-164] Factor effects of plasma or serum concentration and validation of method has been described in the previous section.

Apparatus

Isocratic HPLC equipped with a single channel and a multichannel detector

Nitrogen bomb
Shaker
Light-protected tubes 20 cm^3
Volumetric flask 25 cm^3
Reversed-phase column (such as WATERS Resolve C-18)

Chemicals

Butylated hydroxytoluene (BHT)
Methanol
2 *M* acetate buffer, pH = 5
n-Hexane/dichloromethane/2-propanol mixture 80/19/1 v/v
Retinol, retenyl acetate, retenyl linolate, retenyl palmitate, retenyl
  stearate
Dichloromethane/acetonitrile 20/80 v/v

Procedure

Sample Collection and Storage. In a glass tube containing butylated hydroxytoluene (BHT) the blood samples must be collected and centrifuged at 2000 g for 5 min. The supernatant must be stored in brown tubes with small volume head space above the plasma or serum and stored at –20°C. The tubes must be flushed with nitrogen or argon before use.

Sample Preparation. Add 2 cm^3 methanol to 1 cm^3 serum or plasma and after shaking mechanically; centrifuge at 2000 g for 5 min. Pour the supernatant into a light protected tube (brown) and add 4 cm^3 2 *M* acetate buffer and extract the sample twice with 5 cm^3 n-hexane/dichloromethane/2-propanol 80/19/1 v/v. The organic solvent must be evaporated with nitrogen and add 0.2 cm^3 mobile phase to the residue.

HPLC Conditions. The composition of the eluent is 80/20 v/v dichloromethane/acetonitrile. The detection wavelength for the single-channel UV detector is 325 nm; the wavelength range for multichannel detector is between 230 and 440 nm; the flow rate is 1 cm^3/min.

Validation

Calibration and Linearity. Prepare a stock solution of 100 ng/cm^3 for each analyte in eluent, dilute with eluent to get calibration standards from 1 ng/cm^3 to 100 ng/cm^3 (preferred concentrations are 1, 20, 40, 60, and 100 ng/cm^3). Inject at least twice from each solution.

Selectivity, Specificity, and Interference. Same as described in previous section.

Sensitivity, Limit of Detection, and Limit of Quantitation. Same as described in previous section.

Precision and Accuracy. Extraction recovery and data must be given same as described in previous section.

Evaluation

Calculate the amount of vitamin A content in serum and plasma and give the parameters for validation.

20.5. Other Applications

Determination of Vitamins, Antioxidants, and Other Components in Food and Other Complex Matrices

Theory

Food, foodstuffs, and premixes are complex inhomogeneous mixtures. Isolation and determination of solutes is a difficult task. The measurement can be divided into four steps:

> Sampling (S)
> Homogenization (H)
> Sample preparation to measurement (P)
> Chromatography (C)

Validation of an analytical method means you have to control every step which influences the final results. In our case each step must be validated; it needs much time and labor. The relative standard deviation (RSD) which expresses the total error of analysis is described mathematically as a sum of independent RSDs:

$$\sum RSD = \left[RSD_S^2 + RSD_H^2 + RSD_P^2 + RSD_C^2 \right]^{\frac{1}{2}}$$

If no or little interrelation can be found between the steps and one of these terms is significantly greater than the others, then the total error mainly depends on this step. An attempt to reduce any of the other contributions hardly reduces the total error of the analysis. In an inhomogeneous matrix the sampling is always a very critical step; the lower the analyte concentration, the higher the error. At low analyte concentration the total error can be related to the error of the sampling:

$$\sum RSD \approx \left[RSD_S^2 \right]^{\frac{1}{2}}$$

There are several factors which dictate the sampling such as aim of the analysis, amount of sample analyzed, origin of the sample, concentration level of compounds analyzed, etc. To give general rules for sampling is difficult. There is a general rule that has to be followed: the sample has to be as representative as possible. Some guidelines can be found in the literature.[165-168]

  The second step is the homogenization. Foods have variable texture, structure, viscosity, and phase heterogeneity such as immiscible phases present and so on. Not surprisingly results in the collaborative test show greater coefficient of variation than others. One explanation for this is based on undefined particle size and its distributions and incorrect mixing. Quantifying of reduction of particle size is difficult but necessary; unfortunately, reproducibility of particle size reduction may be poor. The same is true for mixing. Classification, aggregation, and phase separation may prevent adequate mixing. In food analysis sampling and homogenization will be the main sources of error and may affect the accuracy as well as the precision of the analysis. For sample preparation and chromatographic measurement can be used the same rules which are described at other application of chromatographic method.

Determination of Vitamin D in Foods and Feeds

Apparatus

Centrifuge tubes, 25 cm^3 conical centrifuge tubes with glass stoppers
Vortex mixer
Liquid chromatographic equipment with multichannel detector
Chromatographic column, 250 × 4.6 mm, normal phase, silica gel, 5 μm
1.00-cm^3 syringe

Chemicals

Hexane
2-Propanol
Ethyl acetate
Chloroform
Aqueous KOH solution: 400 g KOH in 500 cm^3 water solution
Alcoholic KOH solution: 15 g KOH in 500 cm^3 solution containing 100 cm^3 ethanol
Ethanolic pyrogallol solution: 1 g pyrogallol in 100 cm^3 ethanol
Tartaric acid solution: 4 g tartaric acid in 10 cm^3 water
Mobile phase for HPLC: hexane-2-PrOH, 98:2 v/v
Vitamin D standard stock solution; separately dissolve 50.0 mg each of D_2 and D_3 and dilute to 200 cm^3 with chloroform; working standard solution: take 0.5, 1.0, 2.0, 5.0, and 10.0 cm^3 from stock solution and dilute to 50 cm^3 with hexane; if necessary dilute them further
Saturated NaCl solution

Procedure

Saponification and Extraction.   Place 25 cm^3 milk or 25 g of food or feed in 250 cm^3 glass Erlenmeyer flask; add 50 cm^3 ethanolic pyrogallol solution and 15 cm^3 aqueous KOH solution. Place small magnetic stirring bar in flask and saponify mixture overnight at room temperature with slow stirring. Transfer the mixture to 500 cm^3 separator funnel rinse flask with two 40 cm^3 of saturated NaCl solution and 20 cm^3 of ethanol and 50 cm^3 of hexane separately. Add the mixture to the separator funnel and shake it vigorously. Remove aqueous layer and extract it three times with 50 cm^3 hexane. Combine hexane extracts and wash three times with 60 cm^3 alcoholic KOH solution and wash the extract with 50-cm^3 portion of water until the aqueous layer is base free (no color to phenolphthalein). Concentrate the extract to 5 cm^3 on rotary evaporator, bath temperature below 40°C. Transfer concentrated extract to 25-cm^3 centrifuge tube protected from light and evaporate to nearly dryness (about 50 µl) and dilute it 0.5 cm^3 with hexane; measure the volume of concentrated extract with syringe.

Chromatographic Conditions. Eluent and stationary phases were given above; flow rate is 2 cm^3/min; detection wavelength range is between 220 and 400 nm.

Validation and Evaluation. This must be done as described in Sections 3.3 and 3.6. The amount of D$_2$ and D$_3$ must also be calculated.

Determination of Antibiotic Materials in Food, Foodstuff, Premixes, and Other Matrices

Theory

In the previous section we have driven your attention to specific difficulty in analysis of food and other biological matrices, especially at a low concentration of analyte. The monitoring of food materials for antibiotic residues is an area where chromatographic methods must be used for determination of antibiotic materials at low levels. The problem is further complicated by the wide range of chemical composition of the antibiotic materials. There is no single method of applying determination in food, feed, or other biological matrices. A careful choice of separation method, sample preparation method, and detection method can provide only a reliable result. In literature from the direct injection whole blood, discussed in clinical application), diluting only the sample with eluent (antibiotic residue determination in honey) to very complex solid phase extraction methods can be found.

For the determination of antibiotic residues in milk strong mineral acids have been used for the extraction of the tetracycline group. Deproteinization of tissues must usually be done. If possible deproteinization and extraction must be taken at the same time. Water-miscible organics are largely used for deproteinization and extraction because the procedure is rapid and simple and, with pH adjustment more or less universally applicable, the recovery of analytes is generally high; one portion of extraction liquid can be injected to the chromatographic column or used for further sample preparation.

Determination of Oxytetracyclines in Honey

Apparatus

Isocratic HPLC system equipped with multichannel detector
Vials
Syringes
Chromatographic column: 250 × 4.6 mm Nucleosil C-18, μm

Chemicals

Mobile-phase for OTC determination: 0.01 *M* oxalic acid
containing 0.001 *M* Na₂EDTA-acetonitrile-methanol
70:15:15 v/v
Water
Reversed-phase cartridge for sample preparation

Procedure

Take 1 g of honey and dilute with 5 cm³ of water and add 5 cm³ of eluent, shake it to get a homogenous liquid. From this solution 50 μl aliquot must be injected onto the chromatographic column. If you want to improve the quantitation limit a solid-phase extraction must be applied. Dilute the sample to 30 cm³ with water and prepare a reversed-phase cartridge and force through the sample. Elute the sample 2 cm³ of methanol containing 0.01 *M* acetic acid. Evaporate the methanol with nitrogexolve the residue in eluent (0.5 cm³).

Validation and evaluation must be performed as written in Sections 3.3 and 3.6.

20.6.   Special Applications of HPLC

Separation of Optically Active Compounds

Theory

Enantiomers have identical physical and chemical properties in an isotropic environment except that they rate the plane of polarized light in opposite directions. Racemic mixture which contains equal amounts of enantiomers are not able to rotate the plane of polarized light. Separation of enantiomers is one of the most difficult tasks, because no difference exists between the two enantiomers such as melting point, boiling point, refractive index, spectroscopic properties, and so on. Separation of enantiomers can be divided into two main types: so-called direct and indirect ones. In direct separation either stationary phase or mobile phase (adsorption of optically active additive from mobile phase) must be unisotropic (chiral). Indirect separation is based on the reaction of racemic mixture with a chiral reagent to form a pair of diastereomers. Diastereomers have different physicochemical properties and can be separated with chromatography.

Direct separation of enantiomers based on the three-point rule. According to Pirkle:[169] "Chiral recognition requires a minimum of three simultaneous interactions between the chiral stationary phase (CSP) and at least one of the enantiomers, with at least one of these interactions being stereochemically dependent." This means at least one of the interactions will be absent or will differ significantly to get selectivity to be resolved with a high efficiency column. Many factors may influence the extent of interaction of optical isomers in chiral environment: dipole-dipole, induction of dipole, ion-dipole interactions, electrostatic forces, hydrogen bonding, or hydrophobic interactions. Any difference that exists between the two antipodes formed complexes with chiral agent may be used for their separation, as structural rigidity, steric interferences, solubilities, extent of ionization, ligand formation and so on. Unfortunately these differences are usually very small and capacity factors of two enantiomers differ only with a small value (selectivity of chromatographic system is small or negligible). Changing of chromatographic selectivity can be improved to some extent, but the range where a separation takes place is very narrow. Shifting the conditions to a small extent, no separation can be obtained between the two antipodes. A characteristic feature of enantiomer separation is that there is no general rule, except the three-point rule, which can be used for selecting an appropriate CSP system to solve a particular problem. Even small structural changing in the solute may destroy selectivity. Let us say that to find a CSP is based on more or less trial and error until an acceptable separation method has been found.

Separations of Enantiomers Forming Diastereomers. Resolution of the Enantiomers of the Acidic Moieties of Pyretroids Insecticides

Theory

Enantiomeric mixtures containing a reactive functional group can be derivatized with an optically pure chiral reagent. The products of this reaction are diastereomeric isomers, which can separate in a nonchiral chromatographic method. There are some requirements for these reactions:

> No racemization or epimerization takes place during the reaction
> For all enantiomers there has to be the same conversion rate
> Chiral reagents have to be as pure as possible (no enantiomeric impurities in the reagent)

Some of chiral reagents used for preparation diastereomers are α-naphthyl-(ethyl)isocyanate, 1-(1-phenyl)ethylamine, L-1-(4-dimethylamino-1-naphthyl)-ethylamine, α-methoxy-α-methyl-1-naphthaleneacetic acid, α-methoxy-α-trifluoromethyl-phenylacetyl-chloride, (R)-(+)-α-methoxy-p-nitrobenzylamine, (+)-camphor-10-sulfonyl chloride, and (+)-neomethyl isothiocyanate.

To use indirect separation of enantiomers has advantages and disadvantages. Advantages are easily obtainable; most are easy and fast (the HPLC method development), columns used are relatively cheap, and with a good choice of derivatization reagent the sensitivity of detection can be improved. Disadvantages are frequently time consuming, diastereomers have different UV absorbance, and it is difficult to recover enantiomers.

Apparatus

> Isocratic HPLC system with multichannel UV detector
> Ultrasonic bath
> Rotary evaporator
> Beakers
> Thermostat

Chemicals

> 3-(2,2-Dichlorovinyl)-2,2-dimethylcyclopropanecarbocyclic acid (permethryn acid)
> 3-(2,2-Dimethylvinyl)-2,2-dimethylcyclopropanecarbocyclic acid (chrysantemic acid)
> 3-(2,2-Dibromovinyl)-2,2-dimethylcyclopropanecarbocyclic acid (deltamethryn acid)
> $SOCl_2$
> (–)-1-(1-Phenyl)ethylamine
> Benzene
> Hexane
> Dioxane
> Pyridine
> Silica gel column 5 μm, 250 × 4.6 mm

Procedure

Preparation of Diastereometric Derivatives. Take one of the three acids described at chemicals and put into a vial (5 to 20 mg) and add excess of $SOCl_2$ and put into a thermostat at 50 to 55°C for 2 hours. Remove the unchanged SO_2 by rotary evaporator in vacuum.

Put into the residue (–)-1-(1-phenyl)ethylamine about 1:1.2 motor ratio in dry benzene containing 30 μl of dry pyridine for 70 to 80°C and let it sit at room temperature overnight.

Remove the organic solvent, by rotary evaporator in vacuum, and dissolve the residue in the mobile phase, about 10 cm³ and dilute it if necessary.

Chromatographic Conditions. The mobile phase is a mixture of hexane/dioxane 99/1 v/v. The flow rate is 2 cm³/min; the detection wavelength range is between 215 and 320 nm. Injection volume is 20 μl. Validation of the preparation of diastereometric derivatives and of chromatographic system used has been described in Sections 3.3 and 3.6. Follow that procedure.

Evaluation.   All chromatographic data and validation parameters must be calculated. Follow the procedure given in Sections 3.3 and 3.6.

Separation of Diastereomers by Using Optically Active Stationary Phase

Theory

Several authors have tried to categorize CSP into different classes. They used some criteria for this classification based on the suggested separation mechanism. In our discussion we prefer the following classification:

1.   The solute is part of a diastereometric metal complex. This type of chiral chromatography is called chiral ligand exchange chromatography (LEC).

2.   Charge-transfer CPS. Essentially the only discernible interactions are attractive charge transfer ones.

3.   Pirkle-type CSP or asymmetric strand CSP. The solute-CSP complexes are formed by multiple attractive interactions, including hydrogen bonding, Π-Π, and dipole-dipole interactions between the solute and low molecular weight CSP.

4.   Chiral cavity packing for CSP. Packing classified into this category forms inclusion complex with enantiomers. Main representatives of this CSP are cyclodextrin phases (α, β, γ-cyclodextrins).

5.   Polymer-based packing for CSP. Some of these polymers are derived from cellulose (natural origin) and others are synthetic in origin (polyacrylates and polyacrylamides).

6.   Silica-bound protein packing for CSP. These types of CSP involve the use of proteins as the chiral selector bound on the surface of silica gel. Two proteins are widely used in this class: bovine serum albumin (BSA) and α_1-acid glycoprotein (AGP).

There are several published books and literature about this topic. It is impossible to list all of them. We refer only a few,[170-173] who is interested in separation of optically active solutes for.

Apparatus

Isocratic chromatographic system equipped with a
  UV-detector
Graduated cylinder, 500 cm^3
Beaker
Vacuum evaporator
Magnetic mixer
Weighing machine
Ultrasonic bath
Pirkle-type CSP, (R)-N-(3,5-dinitro-benzoyl-phenyl)-glycine,
  250 × 4.6 mm

Chemicals

Racemic and optically pure norephedrine
Phosgene 12.5% in toluene
Diethyl ether
Ethanol
Hexane
Acetonitrile
10% sodium hydroxide solution
Sodium chloride
Anhydrous sodium sulfate

Procedure

Take racemic and enantiomerically pure norephedrine hydrochloride salts about 100 mg and add 1.5 cm^3 10% sodium hydroxide solution and 5 cm^3 diethyl ether and cool to 0°C in cooling bath containing ice and sodium chloride solution; add 2.5 cm^3 of 12.5% phosgene in toluene dropwise; stir the solution for about 1 hour. Evaporate the organic layer with the rotary evaporator and dry it with anhydrous sodium sulfate; recrystallize the solid from absolute ethanol.

HPLC Conditions. The mobile phase consists of hexane and 2-propanol or ethanol/acetonitrile modifiers. The column is Pirkle type (R)-N-(3,5-dinitro-benzoyl-phenyl)-glycine CSP, 250 × 4.6 mm, d_p = 5 μm. The detector wavelength is about 254 nm; the flow rate is 2 cm^3/min. The separation must be carried out at room temperature.

Validation. The validation procedure has been described in Sections 3.3 and 3.6 and must be followed, except for the ruggedness test of the separation conditions. For this purpose first try the separation with a mobile phase consisting of 97/3 v/v hexane/2-propanol and test the system of the racemic derivative of norephedrine. If the R_s ≥ 1.0 the mobile phase composition is correct; if the R_s < 1.0 and the capacity factor is not higher than 10, decrease the 2-propanol content until R_s ≥ 1.5 and k < 20. If it is not the case, try the separation with a mixture of acetonitrile/ethanol 1/2 v/v adding to 97 v hexane and repeat the procedure as was written for the 2-propanol modifier.

Evaluation

Give all data and plot as written in Sections 3.3 and 3.6 and enantiomeric purity of a commercially available product consisting of (+)norephedrine.

Separation of Optically Active Amino Acids with Adding β-Cyclodextrin (CD) to the Mobile Phase

Theory

CD can be added to the mobile phase and will distribute between the two phases. The resolution and capacity factor of the antipodes affected the absorbed amount of CD, which depends on the eluent concentration of CDs.

Advantages of this technique are it is less expensive, it can be carried out with conventional packing, and concentration of chiral additives can be changed very easily. The selectivity of the chromatographic system differs somewhat from those of CSP packing.

There are several additives used in HPLC to resolve racemic mixture. Some examples will be given below. For separation of racemic amino acids in aqueous eluent metal optically pure amino acid complex can be used. For example, copper(2)-L-proline can be used for separation of underivatized amino acid enantiomers on an ion-exchange column. It is postulated that stereoselectivity can be explained with the complex stability differences in aqueous medium. If the L-proline is changed to D-proline, the elution order will change as well. If the metal complex additives contain hydrophobic segment then the additive will adsorb onto the surface of the reserved-phase material, and retention of the enantiomers can affect the stereoselectivity of both phases.

Some of that hydrophobic ligands are L-2-isopropyl-4-octyl-diethylene-triamine, L-propyl-n-octylamine, L-propyl-n-dodecylamine, *N*-methyl or *N,N*-dimethyl-L-phenylalanine, and so on.

Ion-pair chromatography with chiral ion-pair reagent can sometimes be used for the separation of the enantiomers. If stability of two oppositely charged optically active complexes differ both in the mobile phase and on the stationary phase and they do not counteract, then two antipodes can be separated. (+)-10-Camphor-sulfonic acid, quinine, quinidine, albumin, and tartaric acid derivatives have been tried as chiral ion-pairing reagents. Zwitterions for ion-pairing reagent were also applied as L-leucyl-L-leucyl-L-leucine.

Cyclodextrins and their different derivatives are the most used and studied chiral additives. α-Cyclodextrin has six glucose units (α-CD), β-cyclodextrin has seven (β-CD), and γ-cyclodextrin has eight glucose (γ-CD) units. These three naturally occurring CDs are commercially available. The cavity diameter of α-CD varies from 4.5 to 6.0 Å, for β-CD varies from 6 to 8 Å, and for γ-CD varies from 8 to 10 Å. Since all of the hydroxy groups are on the outside surface of the molecule, the cavity is relatively hydrophobic; consequently, the cavity interacts with the hydrophobic part of the solutes and forms a relatively stabile complex with water-soluble and sparingly soluble molecules. If the stability of the two enantiomer (inclusion complex formation constants differ) inclusion complexes is not the same, there is some chance to separate them. The outer port of CDs can be modified and the solubility of CD interaction with solutes is altered. A huge number of publications can be found about this topic, which is out of the scope of this short summary.

Apparatus

Gradient HPLC equipment with UV detector ($\lambda = 220$ nm)
Graduated cylinder, 500 cm^3

Weighing machine
Ultrasonic bath
Vials
Volumetric flask, 25 cm^3

Chemicals

Dansyl-D,L-amino acids
Amino acids
Acetonitrile
Water
Potassium phosphate
Phosphoric acid
β-Cyclodextrin
Reversed phase column (preferred Hypersil ODS, 5 µm,
   250 × 4.6 mm or 150 × 4.6 mm)

Procedure

Note: if the dansylated amino acids are not in the lab, dansylation of amino acid can be done easily at 40°C for 30 min.

Prepare eluent for isocratic runs with 0 to 15 mM cyclodextrin and contain 20/80 v/v acetonitrile/buffer; the buffer contains 100 mM K$_2$HPO$_4$ and its pH adjusted by 1M H$_3$PO$_4$ to 5.5.

For gradient separation eluent A contain 5/90 v/v acetonitrile/100 mM phosphate buffer (pH = 6.5) and B 25/75 v/v acetonitrile/100 mM phosphate buffer (pH = 5.1), both eluent contain 12.5 mM β-cyclodextrin.

Method Development and Validation

In isocratic runs all cases equilibrate the chromatographic system before any solutes are injected. Equilibrium means there is no retention time shift for dansylated amino acids. Set the detector wavelength at 220 nm and flow rate at 1 cm^3/min. For selection of the appropriate concentration of β-CD, inject dansyl-D,L-asp, or dansyl-D,L-glu. After determination of β-CD concentration where the selectivity factor for solutes chosen the highest value, inject a mixture of dansylated amino acids.

At gradient run start a linear gradient and with a long gradient time of about 140 min of all the dansylated amino acids are present in the mixture. To study the effect of gradient time on separation, change the gradient time and if necessary the shape of gradient (see Section 20.3.). The concentration of dansylated amino acids is about 1.00 nmol/cm^3.

Evaluation

Data and plot which must be given for the validation procedure can be found in Sections 3.3 and 3.6 and must be summarized. Calculate the capacity factors, separation factors, and resolution between the D- and L-dansyl amino acids.

20.7. Using Pre- and Post-Column Reaction to Improve Selectivity and Detection Limit

Theory

General Description of Method

In liquid chromatography the separation can be achieved on the column and the selectivity mainly depends on the right choice of stationary phase and mobile phase. Sensitivity of course affected the column used but the main factor is the detection method used. In practice two types of detector can be chosen. Bulk property and solute property detector can be applied to detect a particular solute. Bulk property detectors are relatively insensitive; a refractive index (RI) detector can detect about micromolar levels of analytes. The solute property detectors offer more sensitive determination of given solutes. UV-VIS detectors give a detection limit about three or four orders lower than RI; fluorescence and electrochemical (EC) detectors give about three orders lower detection limit compared to UV. Detectability of a solute depends on the structure and given functional groups present in a molecule. As not all the substances have UV, fluorescence properties, or cannot be easily oxidized electrochemically, chemical derivatization of labeling must be used to obtain substances for that sensitive, solute property detectors can be used. To achieve this goal two types of labeling reaction can be conducted. Precolumn labeling of solutes can be done off-line prior to the analysis performed and on-line post-column derivatization of analytes can be done in the column elutes. There are several reaction types that can be used for pre-column labeling for UV-VIS, fluorescence, chemiluminescence, and EC detection methods. Reagent can be classified according to properties that are introduced and which functional group can be reacted. Now we refer to general books and review published.[174-177] In precolumn derivatization no limitation in time and temperature used for labeling is determined by some practical requirements. Several samples can be handled simultaneously and reaction can be made in an automatic sampler. Reaction must be as complete as possible or at least the recovery factor must be constant and reproducible. In a pre-column labeling reaction there is no difficulty with the access of the derivatizing reagent.

In a post-column reaction system a second pump (reagent delivery system) must usually be used to deliver the reagent for labeling the solutes separated. The reagent delivery system can be connected before the detector is used. The derivatization can take place in different ways. Some types of reaction do not need a second pump. For example, in a photochemical reaction or in an EC reaction derivatives can be produced which can be monitored by a solute property detector. Another classification of post-column reaction can be done according to the phase or phases where the labeling reactions take place. If the derivatization reactions are performed in a single phase, then this type of reaction is called homogenous. The derivatization can happen in two phases. The second phase can be an immiscible fluid or a solid. There are some publications where the description and application of post-column reaction system can be found.[178-180] Instrumentation of a post-column reaction system will influence the sensitivity and selectivity. To get high sensitivity a pulseless pump must be used, the mixing device has to be a low dead volume, but the mixing can be as perfect as possible. Commercially available mixing devices can work satisfactorily if the viscosity differences between the mixing fluids are relatively small. The labeling reaction can take place in a reactor which has to have a given volume. This dead volume causes significant extra column broadening. There are three ways to reduce this effect: coil the tubing, knit or weave the tubing, or pack it with solid particles (bed reactor). The reactor can be placed in some sort of heating device to increase the reaction rate. There are some dedicated applications of post-column reaction: determination of amino acids, carbohydrates, carbamate pesticides, transition metals, and acetylcholine. The radiochemical detection and detection in chemically suppressed ion chromatography are also post-column reactor systems.

Determination of Amino Acids Using Pre-Column Fluorescence Derivatization

Theory

Amino acid analysis is widely used in biochemistry. Determination of primary structure of proteins and peptide sequencing all require a sensitive and selective method for amino acid analysis. For the analysis of protein structure possessing only small amounts of sample are available. Amino acids processing no significant chromophores or fluorophores in addition to this in a complex mixture large number of substance may be interfere with amino acids. Amino acid analyzers were developed based on ion exchange and operated in post-column derivatization mode utilized either ninhidrin or other fluorogenic reagents such as o-phthalaldehyde. These instruments have been used in different fields, but they have some shortcomings: high cost of instrumentation, long analysis time, and inadequate separation and detection limit for some solutes such as proline, hydroxyproline, cysteine, and cystine. Using HPLC most of the shortcomings can be overcome. It has been demonstrated that o-phthalaldehyde in the presence of either ethanethiol or mercaptoethanol can react with amino acids rapidly.

Apparatus

> Microprocessor or PC-controlled binary gradient system
> Fluorescence detector
> Block thermostat
> Ultrasonic bath
> Nitrogen bomb

Ignition tubes
Weighing machine
Graduated cylinder 500 cm^3
Hamilton micro-syringe
Vortex mixer
Capped vials (1, 5, 25 cm^3)

Chemicals

o-Phthalaldehyde (OPA)
Mercaptoethanol
0.4 *M* borate buffer (pH = 9.5, adjusted by 4 *M* NaOH)
4 *M* methanesulfonic acid containing 0.2% 3-(2-aminoethyl)
 indole
2-Propanol
Solid carbon dioxide
Methanol
Acetonitrile
Tetrahydrofuran
High purity water
Amino acid standard
Sodium acetate
Sodium dodecylsulfate
Amino ethanol
30% hydrogen peroxide
Formic acid
4-Chloro-7-nitrobenzofurazan (NBD)
Insulin

Procedure

Hydrolysis and Derivatization.   Take about 100 μg of the protein (insulin) sample into an ignition tube and add 200 μl of 4 *M* methanesulfonic acid
containing 0.2% of 3-(2-aminoethyl)indole and freeze it in a slurry of carbon
dioxide and isopropanol. Dissolved gases must be removed by evaporating the
tube by freezing-thawing twice. Seal the tube under vacuum and place a block
thermostat at 110°C for 24 hours. After the acidic hydrolysis neutralize the
sample with 4 *M* NaOH. If the concentration of amino acids is much higher
than 25 nmol/cm^3 dilute the sample with 0.4 *M* borate buffer (pH = 9.5). Take
an aliquot with a Hamilton micro-syringe and mix with an equal volume of 4
μg/cm^3 amino-ethanol solution as an internal standard. Take one part from this
solution and four parts of 1% (w/v) sodium-dodecylate in 0.4 *M* borate buffer
and four parts of OPA solution and mix vigorously for 1 min using a Vortex
mixer and inject 10 μl onto the chromatographic column immediately.

The procedure is the same for amino acids except for the secondary amino
acids such as proline and hydroxyproline. Cysteine and cystine must be oxidized

before pre-column derivatization. Amino acid standards can be bought in a mixture or separately. The concentration for pre-column separation is about 25 nmol/cm^3 separately.

Performic Acid Oxidation.   Freeze 100 μg of protein sample in drug ice and lyophilized acid and let it stand in a closed container for 2 h at ambient temperature, cool it to 0°C, and add 100 μl to the sample and allowed it to stand in a capped vial for 2.5 h at 0°C; add 0.9 cm^3 of cold water, then lyophilize 200 μl from this mixture. The hydrolysis and derivatization with OPA must be done as described before.

Preparation and Storage of OPA.   Weigh 50 mg of OPA in a capped vial and dissolve in 1.5 cm^3 of methanol; add 50 μl mercaptoethanol and 11 cm^3 of 0.4 M borate buffer (pH = 9.5), mix the solution and flush with nitrogen to remove the oxygen. The solution must be stored in the dark and allowed to stand for 24 h before use. Every 2 days 10 μl of mercaptoethanol must be added to it. The solution can be used about 2 weeks.

Derivatization of Proline and Hydroxyproline.   The same hydrolization must be used as prepared for OPA derivatization. An aliquot of the sample must be taken and neutralized with 4 M NaOH; add an equal volume of NDB solution (c = 2 mg/cm^3) in borate buffer, heat it in a capped vial to 60°C for 5 min, cool the mixture to 0°C, and inject 20 μl onto the chromatographic column.

Chromatographic Conditions.   For separation a reversed-phase column (C-18, ODS) can be used with particle diameter of about 5 μm, the preferred column dimension are 250 × 4.6 mm and stationary phase is Ultrasphere ODS. The excitation wavelength is 330 nm for OPA derivatives and measuring (emission) wavelength is 420 nm, respectively. The flow rate is 1.0 cm^3/min.; the gradient program and eluent composition for OPA derivatives are

Solvent A: THF-0.05 M sodium acetate, pH = 6.6, 1:99 v/v
Solvent B: methanol/water 97/3 v/v

and use a linear program and gradient time is 60 min for OPA derivatives and for NBD derivatives the program start 30/70 v/v A/B to 45/55 v/v A/B gradient run time is 10 min.

Validation of Method
Selectivity.   Prepare the ODA and DNB derivatives of amino acids described above and perform the analysis parameter given above. If all the solutes resolve the gradient steepness and gradient run time is good for determination. If some of derivatives coelute change the gradient steepness and gradient run time until all components will resolve (R > 1.0).

Linearity and Range.   Prepare derivatives with the concentration range 100 pmol/cm^3 and 100 nmol/cm^3 range for at least five different concentrations. Inject at least twice onto the chromatographic column.

Limit of Detection.   Prepare derivatives with the concentration at about 10 to 100 pmol/cm³, inject onto the chromatographic column, and estimate which solute concentrations give three times higher signal than the noise; prepare them and inject onto the column three times.

Limit of Quantitation.   Prepare at least five times higher concentration from standard amino acids than the limit of detection; at least five independent measurements must be done.

Accuracy (Recovery).   Take a known protein (amino acid content certificate); at least five independent determinations must be performed.

Precision.   Take five aliquots from hydrolyzed protein and prepare the derivatives; inject at least twice onto the chromatographic column.

Ruggedness Test, Reproducibility.   Prepare the derivatives five times a day on five different days and measure them.

Ruggedness Testing of Column.   Change the column to another which was supplied by the same firm and change the columns at least two of which were supplied by different firms but chemically equivalent with used before (C-18 or ODS), and carried out of measurement with derivatives.

Ruggedness Testing of Mobile Phase Composition.   Change the A eluent pH between 5 and 7.5 and repeat the separation; change the buffer concentration between 0.04 and 0.06 M and conduct the measurement. Change the B eluent concentration to 95/5 v/v methanol/water and repeat the measurement. Change the gradient time with ±10% and repeat the measurement.

Ruggedness Test of Sample Preparation.   Vary the amount of protein between 80 and 120 mg and measure the amino acid derivatives. Change the borate buffer and sodium dodecylsulfate concentration ±10% and repeat all steps. Vary the buffer pH between 9 and 10 and repeat the measurement. Vary the reaction time with ±10% and repeat the measurement.

Evaluation.   Give the retention parameters and their standard deviation for amino acid derivatives. Calculate the sensitivity, the limit of detection, the limit of quantitation, the linearity and range, accuracy (recovery), and day-to-day and day-in reproducibility. Give data for ruggedness testing of column, eluent, and sample preparation.

Determination of Fatty Acids with Pre-Column Derivatization

Theory

The separation of long-chain fatty acid mixtures has been applied to obtain information of biological samples. Gas chromatography (GC) is frequently used for analysis of methyl, benzyl, and pentafluorobenzyl derivatives. If the fatty acid is a polyunsaturated one, some degradation might happen when the solute is separated by GC. The fatty acids with no conjugated double bonds have only small UV-light absorbtivity. The preparation of UV-absorbing derivatives has been essential to obtain the sensitivity in the nanogram range. One of the most frequently used derivatization methods is the reaction between phenacyl bromide and fatty and carbocyclic acid:

$$RCOOH + BrCH_2COC_6H_5 \rightarrow RCOOCH_2COC_6H_5$$

The reaction takes place in the presence of a catalyst. The suggested catalysts are triethylamine,[181] fluoride,[182] and 18-crown-6.[183] It has been stated that 18-crown-6 proved to be very effective for the derivatization of carbocyclic acid[184] but another author has used triethylamine in acetone with satisfactory results.[181]

Apparatus

Gradient HPLC system coupled to a UV-(VIS) detector and a multichannel detector PDA

Graduated cylinder 500 cm^3

Brown capped vials

Ultrasonic bath

Block thermostat

Reversed-phase column (RP-18, C-18, ODS, preferred is Spherisorb ODS, 250 × 4.6 mm)

Chemicals

Phenacyl bromide reagent with 18-crown-6 catalisator

Acetone

Water

Acetonitrile

Triethylamine

cis-5,8,11,14,17-Eicosa-pentaneoic acid (EPA)

cis-4,7,13,16,19-Docosa-hexaneoic acid (DHA)

Arachidonic acid

Linolenic acid

Oleic acid

Methanol

Sodium hydroxide

Stearic acid

Palmitic acid

Procedure

Derivatization. Take approximately 0.1 mg of fatty acids and 20 µl of phenacyl bromide solution (10 mg/cm^3) and 20 µl of triethylamine solution (10 mg/cm^3) 2 cm^3 in acetone, mix them, and store at 50°C for 2 hours; 20 µl of this solution must be injected onto the chromatographic column.

Hydrolysis of Fish Oil.   Take about 2 mg of fish oil into a capped brown vial containing 2 cm^3 alkaline methanol, purge with nitrogen to remove the dissolved oxygen, and put it into a block thermostat for 2 hours. Put from the reaction mixture about 1 cm^3 into a brown vial and remove the methanol with nitrogen. From the dried sample the derivatization must be done as described above.

Chromatographic Conditions.   Reversed-phase column (C-18, ODS, RP-18) must be used

> Eluent A: 60/40 v/v acetonitrile/water
> Eluent B: 97.5/2.5 v/v acetonitrile/water

Detector wavelength is 254 nm; flow rate is 1 cm^3/min; gradient time is 38 min.

Validation

Selectivity, Specificity, and Interference.   Take a standard mixture containing fatty acids listed above and after derivatization make a run at the chromatographic conditions given above. If all the solutes are resolved ($R_s \geq$ 1.5), take a fish oil sample and perform a chromatographic run. Compare the retention time, peak width, and ratio plot of all chromatographic peaks and if they are pure (no interference, peak purity, and homogenity investigation must be done by multichannel detector PDA are used and has been written before) the chromatographic conditions are correct to run a real sample. In another case, change the gradient time and if necessary the composition of A and B solvent until the chromatographic run meets the demand of resolution and peak purity criteria. Make a blank sample and run it.

Calibration and Linearity.   Take 0.01, 0.05, 0.1, 0.2, and 5 mg of EPA and DHA and derivatize them according to the description that has been given above and keep the molar ratio constant. Inject from each solution at least twice onto the chromatographic column.

Limit of Detection (LOD).   Used the solution prepared above and dilute until the signal/noise ratio is about 3. At this concentration inject three times.

Limit of Quantitation (LOQ).   Use a solution which has the concentration about five times compared to LOD and take at least five different samples; at each concentration inject at least twice.

Recovery (Accuracy).   Take a certificate oil sample containing EPA and DHA or other fatty acids and prepare at least five independent samples; after sample preparation inject onto the chromatographic column.

Precision and Reproducibility.   Take a sample and divide it into ten different parts and make five sample preparations in 1 day, each day of the week and inject at least twice onto the chromatographic column.

Validation of Pre-Column Derivatization.   Variables that must be changed to investigate the ruggedness of the sample preparation are the following:

Solvent volume of acetone between 1.5 and 2.2 cm^3
Temperature between 55 and 65°C
Time for reaction between 1.6 and 2.4 hours
Change the molar ratio with ±20% applied above
Investigate the stability of derivatives at different temperatures
    and different times

Evaluation

Give the linearity, sensitivity, detection limit, limit of quantitation, recovery, (accuracy), day-in reproducibility, day-to-day reproducibility, and ruggedness of sample preparation for the parameters given above. Calculate the EPA and DHA content of the sample and its relative standard deviation for at least five different measurements.

Determination of Polyamines

Theory

The naturally occurring polyamines (putrescence, cadaverine, spermidine, and spermine) are present in all living cells. Elevated levels of intracellular polyamines are correlated with increased rates of cellular proliferation and might be associated with several different pathological conditions. Polyamines have low UV absorbance, no fluorescence activity, cannot be detected by EC detection (amperometric), and the concentration is not sufficient to detect with RI in tissues and body fluids or in any other matrices. There are several reagents for pre-column derivatization. With o-phthalaldehyde-2-mercapto-ethanol or fluorescamine or other reagents have been published in.[58] Both have disadvantages, even the stability of derivatized compound or the lengthy and laborious derivatization procedure and so on. Derivatization of polyamines with benzoyl chloride has been accepted to get stable and UV active compounds.

Apparatus

Isocratic HPLC equipped with UV detector and multichannel
    detector in series
Centrifuge
Ultrasonic bath
Screw-capped plastic vials 10 cm
Graduated cylinder 500 cm^3
Volumetric flasks 25 cm^3
Vortex mixer
Reversed-phase column, C-18 or ODS (preferred is a DB
    column as Supelcosil LC-18- DB 5 μm)

Chemicals

> Putrescine
> Cadaverine
> Spermidine
> Spermine
> Methanol
> 2% benzoyl chloride in methanol solution
> 5% perchloric acid
> Water
> Saturated sodium chloride solution
> 2 *M* sodium hydroxide
> Membrane filter device (Millipore HU filters) 0.45 μm

Procedure

Derivatization.   Stock solution from each polyamine must be prepared about 200 ng/cm^3; dilute with eluent and make solutions containing 1, 20, 40, 60, and 100 ng/cm^3 for each of the solutes. Take 1 cm^3 from each solution and add 1 cm^3 stock solution of 2% benzoyl chloride in methanol and 1 cm^3 of 2 *M* sodium hydroxide; shake the mixture with Vortex mixer and incubate at 37°C for 20 min in screw-capped plastic tubes. Add 3 cm^3 of diethyl ether and 2 cm^3 of saturated sodium chloride solution, shake with Vortex mixer about 1 or 2 min, then centrifuge it at 3000 g for 10 min to separate the aqueous phase and organic one. Transfer the upper organic solvent phase into another screw-capped plastic tube and evaporate with nitrogen to dryness. Dissolve the residue in 0.2 cm^3 methanol and shake the solution with Vortex. With a sample, e.g., tissues of mouse or other biological fluids, the solution must be filtered with a membrane filter. The dissolved samples in methanol can be stored at –20°C for about 2 weeks. The residue can be stored at –20°C for more than 1 month.

Sample Preparation.   Take 200 mg of mouse brain and kidney and add 10 cm^3 of 5% perchloric acid, sonicate it 3 min, and centrifuge about 20,000 g for 15 min at 4°C. Remove the supernatant into plastic vials. These solutions must be used for derivatization described above.

Chromatographic Conditions.   For separation of derivatized polyamines a deactivated reversed-phase column must be preferred. The separation can be performed with any C-18 or ODS columns. The eluent is 60/40 v/v methanol/ water mixture, detection wavelength is 254 nm, and flow rate 1 cm^3/min.

Validation.   All parameters can be checked as described in previous paragraphs. In addition the effect of sonication time on recovery must be checked as well.

Evaluation.   All data and plots must be given which are described in previous paragraphs.

20.8. Validation

According to Szepesi[185] validation is a process of evaluation which will demonstrate the method that is scientifically sound under experimental conditions intended for use. From this definition it can be concluded that there is no general routine of evaluation for different method, different sample preparation, and so on, but the old chemical analytical role must be applied. It says you have to investigate every parameter and condition which influences the result of an analytical procedure. You have to focus your attention to the "narrow neck" effect, which will influence mainly the real value of a selected analytical result.

In practical application of HPLC we have described the procedure which has to be done for method validation which include the system validation procedure as well. For example in environmental analysis one of the main sources of error is sampling (see Section 20.4.2); the same is true for the application of HPLC in food analysis (see Section 20.5). The procedure is quite time consuming but working with an unvalidated method makes the results useless. Confirmation of results in each application of HPLC is badly needed. It can be done by applying independent methods of PAH determination with HPLC and GC-MS and using coupled detectors, e.g., multiwavelength (PDA) and electrochemical detector (example given in Section 20.3.3).

Most of the cases sample preparation must be done before the analytical measurement. It can be seen from the published methods that this step is critical and will determine the final results. One must keep in mind that HPLC technology has been improved greatly in the past 10 years and is highly automated. Sample preparation will depend on matrices and solutes analyzed; for automating the sample preparation is still in the developing stage. Knowledge of analytical chemistry cannot be substituted by a automatic sample processor even with highly sophisticated software. If somebody wants to practice the validation we refer to Sections 20.4.2 and 20.7, where we try to write down validated methods, but if somebody wants to get even more detail it can be found in Szepesi,[185] and in two review articles.[186,187] Validation procedures can be done in each of the practical examples given in Section 20.6.

20.9. References

1. **Bristow, P. A. and Knox, J. H.,** *Chromatography,* 10, 279, 1977.

2. **Foley, I. P. and Dorsey, I. G.,** *Anal. Chem.,* 55, 730, 1983.

3. **Wilkie, C. R. and Chang, P.,** *Am. Inst. Chem. Eng.,* 1, 264, 1955.

4. **Huss, V., Chevalier, I. L., and Siouffi, A. M.,** *J. Chromatogr.,* 500, 241, 1990.

5. **Unger, K. K.,** *Porous Silica,* Elsevier, Amsterdam, 1979, p. 183.

6. **Poole, C. F. and Poole, S. K.,** *Chromatography Today,* Elsevier, Amsterdam, 1991, p. 362.

7. **Van den Riest, P. I., Riteic, H. I., and Rose S.,** *LC-GC,* 6, 124, 1988.

8. **Walters, M. I.,** *J. Assoc. Off. Anal. Chem.,* 70, 465, 1987.

9. **Kimata, K., Iwaguchi, K., Onishi, S., Jinno, K., Eksteen, R., Hosoya, K., Araki, M., and Tanoka, N.,** *J. Chrom. Sci.,* 27, 721, 1989.

10. **Engelhardt, H. and Junghcim, M.,** *Chromatography,* 29, 59, 1990.

11. **Nant, C. T. and Hodes, S.,** *Chromatographia,* 24, 805, 1987.

12. **Sander, L. C. and Wise, S. A.,** *Crit. Rev. Anal. Chem.,* 18, 299, 1987.

13. **Sander, L. C. and Wise, S. A.,** *HRC CC,* 11, 383, 1988.

14. **Sander, L. C. and Wise, S. A.,** *LC-GC,* 8, 387, 1990.

15. **Sander, L. C.,** *J. Chrom. Sci.,* 26, 380, 1988.

16. **Sander, L. C. and Wise, S. A.,** *Adv. Chrom.,* 25, 139, 1986.

17. **Weber, S. G. and Tramposh, W. O.,** *Anal. Chem.,* 55, 1771, 1983.

18. **Cox, C. B. and Stuat, R. W.,** *J. Chrom.,* 384, 315, 1987.

19. **Kohler, I. and Kirkland, I. I.,** *J. Chrom.,* 385, 125, 1987.

20. **Nawrocki, I., Moir, D. L., and Szczepaniak, W.,** *Chromatographia,* 28, 143, 1989.

21. **Scott, R. P. W. and Kucera, P.,** *J. Chrom.,* 142, 213, 1977.

22. **Nondek, L., Buszewski, B., and Berek, D.,** *J. Chrom.,* 360, 241, 1986.

23. **Engelhardt, H., Dreyer, D., and Shmidt, H.,** *Chromatographia,* 16, 11, 1982.

24. **Verzele, M. and Dewaele, C.,** *Chromatographia,* 18, 84, 1984.

25. **Engelhardt, H. and Jungheim, M.,** *Chromatographia,* 29, 59, 1990.

26. **Sadek, D. C. and Carr, P. W.,** *J. Chrom. Sci.,* 21, 314, 1983.

27. **Kimata, K., Iwaguchi, I., Onishi, S., Jinno, K., Eksten, R., Hosoya, K., Araki, M., and Tanaka, N.,** *J. Chrom. Sci.,* 27, 721, 1989.

28. **Sadek, P. C., Koester, C. I., and Bowers, L. D.,** *J. Chrom. Sci.,* 25, 489, 1987.

29. **Bjerrum, N.,** *Kgt. Danske Vidensk. Selsk.,* 7, 9, 1926.

30. **Denison, I. T. and Ramsey, I. B.,** *J. Am. Chem. Soc.,* 77, 2615, 1955.

31. **Assascina, F., D'Aprano, A., and Fouss, R. M.,** *J. Am. Chem. Soc.,* 81, 1059, 1959.

32. **Szwarc, M.,** *Ions and Ion-pairs in Organic Reactions,* Vol. 1, Wiley-Interscience, New York, 1972, chap. 1.

33. **Sadek, H. and Fouss, R. M.,** *J. Am. Chem. Soc.,* 76, 5897 and 5905, 1954.

34. **Winstein, S., Cliooinger, E., Fainberg, A. H., and Robinson, G. C.,** *J. Am. Chem. Soc.,* 76, 2597, 1954.

35. **Fouss, R. M.,** *J. Phys. Chem.,* 82, 2427, 1978.

36. **Shill, G.,** in *Ion Exchange and Solvent Extraction,* Vol. 6, Mirnsky, I. A. and Marcus, Y., Eds., Marcel Dekker, New York, 1974, 1.

37. **Shill, G., Ehrsson, H., Vessman, I., and Westerlund, D.,** *Separation Methods for Drugs and Related Organic Compounds,* Swedish Pharmaceutical Press, Stockholm, 1983.

38. **Horváth, Cs., Melander, W., Molnár, I., and Molnár, P.,** *Anal. Chem.,* 49, 2295, 1977.

39. **Shill, G., Modin, R., Borg, K. O., and Persson, B. A.,** in *Drug Fate and Metabolism, Methods and Techniques,* Vol. 1, Garret, E. R. and Hirtz, I. C., Eds., Marcel Dekker, New York, 1977.

40. **Bartha, A., Vigh, G., Billiet, H. A., and De Golan, L.,** *J. Chrom.,* 303, 29, 1984.

41. **Gennoro, M. C.,** *J. Chrom.,* 449, 103, 1988.

42. **Bartha, A., Vigh, G., and Varga-Puchony, Z.,** *J. Chrom.,* 499, 423, 1990.

43. **Goldberg, A. P., Nowakowsha, E., Antle, P. E., and Snyder, L. R.,** *J. Chrom.,* 316, 241, 1984.

44. **Billiet, H. A., Vuik, I., Strasters, I. K., and De Golan, L.,** *J. Chrom.,* 384, 153, 1987.

45. **Coenegracht, P. M. I., Van Tuyen, N., Metting, H. I., and Coenegracht-Lamers, P. M. I.,** *J. Chrom.,* 389, 351, 1987.

46. **Crombeen, I. P., Kreak, I. C., and Poppe, H.,** *J. Chrom.,* 167, 219, 1978.

47. **Bartha, A. and Vigh, G.,** *J. Chrom.,* 260, 337, 1980.

48. **Juergens, V.,** *J. Liq. Chromatogr.,* 11, 1925, 1988.

49. **Hearn, M. T. W., Ed.,** *Ion-Pair Chromatography: Theory and Biological and Pharmaceutical Applications,* Marcel Dekker, New York, 1985.

50. **Gennaro, M. C.,** *J. Chrom.,* 449, 103, 1988.

51. **Knox, I. H. and Laird, G.,** *J. Chrom.,* 125, 89, 1976.

52. **Terwcij-Groen, C. P., Heemstra, S., and Kraak, I. C.,** *J. Chrom.,* 161, 69, 1978.

53. **Scott, R. P. W. and Kucera, P.,** *J. Chrom.,* 175, 51, 1979.

54. **Bildingmeyer, B. A.,** *J. Chrom. Sci.,* 18, 525, 1980.

55. **Billiet, H. A., Vuik, I., Strasters, I. K., and De Golan, L.,** *J. Chrom. Sci.,* 384, 29, 1987.

56. **Bartha, A., Billiet, H. A., De Golan, L., and Vigh, Gy.,** *J. Chrom.,* 291, 91, 1984.

57. **Tomilson, E., Riley, C. M., and Jefferies, T. M.,** *J. Chrom.,* 173, 89, 1979.

58. **Tomilson, E., Jefferies, T. M., and Riley, C. M.,** *J. Chrom.,* 159, 315, 1978.

59. **Bildingmayer, B. A., Deming, S. N., Price, W. P., Sachoh, B., and Petrusek, M.,** *J. Chrom.,* 186, 419, 1979.

60. **Stahlberg, J.,** *J. Chrom.,* 356, 231, 1986.

61. **Stahlberg, J.,** *J. Chrom.,* 24, 820, 1987.

62. **Stahlberg, J. and Bartha, A.,** *J. Chrom.,* 456, 253, 1988.

63. **Stahlberg, J. and Hagglund, I.,** *Anal. Chem.,* 60, 1958, 1988.

64. **Bartha, A., Vigh, G., and Stahlberg, J.,** *J. Chrom.,* 506, 85, 1990.

65. **Small, H., Stevens, T., and Bauman, W.,** *Anal. Chem.,* 47, 1801, 1975.

66. **Phol, C.,** Chromatographic Separation and Quantitative Analysis of Ionic Species, U.S. Patent, 4, 265, 634.

67. **Rich, W., Johnson, E., and Sidelbottom, T.,** Method and Apparatus for Quantitative Analysis of Weekly Ionized Anions, U.S. Patent, 4,242,097.

68. **Fritz, I. S., Gjeerde, D. T., and Pohlandt, C.,** *Ion Chromatography,* Hueting, New York, 1982.

69. **Skelly, N. E.,** *Anal. Chem.,* 54, 712, 1982.

70. **de Kleijn, I. P.,** *Analyst,* 107, 223, 1982.

71. **Siergeij, R. W. and Danielson, N. D.,** *J. Chrom. Sci.,* 21, 362, 1983.

72. **Hughes, S., Meschi, P. L., and Johnson, D. C.,** *Anal. Chim. Acta,* 132, 1, 1981.

73. **Sherwood, G. A. and Johnson, D. C.,** *Anal. Chim. Acta,* 129, 101, 1981.

74. **Imanari, T., Ogata, K., and Tanabe, S.,** *Chem. Pharm. Bull.,* 30, 374, 1982.

75. **Haddad, P. R., Alexander, P. W., and Trojanowicz, M.,** *J. Chrom.,* 321, 363, 1985.

76. **Deguchi, T., Kuma, T., and Nagai, H.,** *J. Chrom.,* 152, 349, 1978.

77. **Buytenhujs, F. A.,** *J. Chrom.,* 218, 57, 1981.

78. **Yoza, N., Ito, K., Harai, Y., and Ohashi, S.,** *J. Chrom.,* 196, 471, 1980.

79. **Bly, D. D.,** in *Physical Methods in Macromolecular Chemistry,* Vol. 2, Caroll, B., Ed., Marcel Dekker, New York, 1972, 1.

80. **Altgelt, H.,** *Adv. Chromatogr.,* 7, 3, 1978.

81. **Flodin, P.,** Thesis, University of Uppsala, Sweden, 1962.

82. **Ackers, G. K.,** *Biochemistry,* 3, 723, 1964.

83. **Yau, W. W. and Malone, C. P.,** *J. Polym. Sci. B,* 5, 663, 1967.

84. **Yau, W. W.,** *J. Polym. Sci. A-2,* 7, 483, 1969.

85. **Casassa, E. F.,** *J. Phys. Chem.,* 75, 3929, 1971.

86. **Casassa, E. F.,** *J. Polym. Sci. B,* 5, 773, 1967.

87. **Giddings, I. C., Kucera, E., Russel, C. P., and Myers, M. M.,** *J. Phys. Chem.,* 72, 4397, 1986.

88. **Billmeyer, W. and Altgelt, K. H.,** in *Gel Permeation Chromatography,* Altgelt, K. H. and Segal, L., Eds., Marcel Dekker, New York, 1971, 3.

89. **Yau, W. W., Kirkland, I. I., and Bly, D. D.,** *Modern Size-Exclusion Liquid Chromatography,* John Wiley & Sons, New York, 1979.

90. **Ionca, I., Ed.,** *Steric Exclusion Liquid Chromatography of Polymers,* Marcel Dekker, New York, 1984.

91. **Cooper, A. R., Ed.,** *Determination of Molecular Weight,* John Wiley & Sons, New York, 1989, 263.

92. **Spatorico, A. L. and Beyer, G. L.,** *J. Appl. Polym. Sci.,* 19, 1601, 1974.

93. **Spatorico, A. L. and Beyer, G. L.,** *J. Appl. Polym. Sci. B,* 19, 2933, 1973.

94. **Otocka, E. P. and Hellman, M. Y.,** *J. Polym. Sci. B,* 12, 331, 1974.

95. **Prouder, T., Ed.,** *Detection and Data Analysis in Size Exclusion Chromatography,* American Chemical Society, Washington, DC, 1987.

96. **Coulombe, S.,** *J. Chrom. Sci.,* 26, 1, 1988.

97. **Takagi, T.,** *J. Chrom.,* 506, 409, 1990.

98. **Krull, I. S., Mhatre, R., and Stuting, H. H.,** *Trends Anal. Chem.,* 8, 260, 1989.

99. **Unger, K. K.,** *Porous Silica,* Vol. 16, Elsevier, Amsterdam, 1979, 271.

100. **Snyder, L. R., Dolan, I. W., and Grant, I. R.,** *J. Chrom.,* 165, 3 and 31, 1979.

101. **Snyder, L. R.,** in *High Performance Liquid Chromatography, Advances and Perspectives,* Vol. 1, Horvath, Cs., Ed., Academic Press, New York, 1980, 207.

102. Snyder, L. R., Stadalius, M. A., and Quarry, M. A., *Anal. Chem.,* 55, 1412.A, 1983.

103. Dolan, I. W., Lammen, D. C., and Snyder, L. R., *J. Chrom.,* 485, 91, 1989.

104. Stadalius, M. A., Gold, H. S., and Snyder, L. R., *J. Chrom.,* 296, 31, 1984.

105. Quarry, M. A., Grob, R. L., and Snyder, L. R., *Anal. Chem.,* 58, 907, 1986.

106. Quarry, M. A., Grob, R. L., Snyder, L. R., Dolan, I. W., and Rigney, M. P., *J. Chrom.,* 384, 163, 1987.

107. Christ, B. D. F., Cooperman, B. S., and Snyder, L. R., *J. Chrom.,* 459, 1, 1988.

108. Christ, B. D. F., Cooperman, B. S., and Snyder, L. R., *J. Chrom.,* 459, 25, 1988.

109. Christ, B. D. F., Cooperman, B. S., and Snyder, L. R., *J. Chrom.,* 459, 43, 1988.

110. Schoenmakers, P. I., *Optimization of Chromatographic Selectivity,* (Journal of Chromatography Library, Vol. 35), Elsevier, Amsterdam, 1986.

111. Poole, C. H. and Poole, S. K., *Chromatography Today,* Elsevier, 1991, 458.

112. Snyder, L. R., Golach, I. L., and Kirkland, I. I., *Practical HPLC Method Development,* John Wiley & Sons, New York, 1988.

113. Quarry, M. A., Grob, R. L., and Snyder, L. R., *Anal. Chem.,* 58, 907, 1986.

114. Snyder, L. R. and Quarry, M. A., *J. Liq. Chromatogr.,* 10, 17989, 1987.

115. Quarry, M. A., Grob, R. L., Snyder, L. R., Dolan, I. W., and Rigney, M. P., *J. Chrom.,* 384, 163, 1987.

116. Snyder, L. R., in *HPLC, Advances and Perspectives,* Vol. 1, Horvath, Cs., Ed., Academic Press, New York, 1980, 207.

117. Dolan, I. W., Snyder, L. R., and Quarry, M. A., *Chromatographia,* 24, 261, 1987.

118. Eslami, M., Stuart, I. D., and Cohen, K. A., *J. Chrom.,* 411, 121, 1987.

119. Stuart, I. D., Lisi, D. D., and Snyder, L. R., *J. Chrom.,* 485, 657, 1989.

120. Berridge, J. C., *The Techniques for Automated Optimization of HPLC Separation,* John Wiley & Sons, New York, 1985.

121. Glajch, J. L. and Snyder, L. R., Eds., *Computer Assisted Method Development for High Performance Liquid Chromatography,* Elsevier, Amsterdam, 1990.

122. Golshan-Shirazi, G. and Guichon, G., *Anal. Chem.,* 61, 1276, 1989.

123. Guiochon, G. and Kate, A., *Chromatographia,* 24, 165, 1987.

124. **Jones, K.,** *Chromatographia,* 25, 547, 1988.

125. **Tarter, G., Ed.,** *Ion Chromatography,* (Chrom. Sci. Vol. 37), Marcel Dekker, New York, 1987, 12.

126. **Frei, R. W. and Zech, K., Eds.,** *Selective Sample Handling and Detection in High-Performance Liquid Chromatography,* (Journal of Chromatography Library, Vol. 39A), Elsevier, New York, 1988, 145.

127. **Symons, R. K. and Crick, I.,** *Anal. Chim. Acta,* 151, 237, 1983.

128. **Von Noort, P. C. M. and Wondergem, E.,** *J. Chrom.,* 172, 335, 1982.

129. **Kicinski, H. G., Adamek, S., and Kettrup, A.,** *Chromatographia,* 28, 203, 1989.

130. **Ogan, K., Katz, E., and Slavin, W.,** *J. Chrom. Sci.,* 16, 517, 1987.

131. **Eisenbeiss, F., Hein, H., Ioester, R., and Naundorf, G.,** *Chromatogr. Newsl.,* 6, 8, 1978.

132. **Oyler, A. R., Bodenner, D. L., Weich, K. I., Livkhonen, R. I., Carison, R. M., Kopperman, H. L., and Capie, R.,** *Anal. Chem.,* 50, 837, 1978.

133. **Poole, C. F. and Schuetle, S. A.,** *J. High Resol. Chromatogr. Commun.,* 6, 526, 1983.

134. **Chadek, E. and Marano, R. S.,** *J. Chrom. Sci.,* 22, 313, 1984.

135. **Ghaoui, L.,** *J. Chrom.,* 399, 69, 1987.

136. **Andelman, I. P. and Snodgrass, I. E.,** *Crit. Rev. Environ. Control,* January, 69, 1974.

137. **ACS Committee on Environmental Improvement,** *Anal. Chem.,* 52, 2242, 1980.

138. **Kissinger, P. T., Refshavge, C., Dreiling, R., and Adams, R. N.,** *Anal. Lett.,* 6, 465, 1978.

139. **Stulik, K. and Pacakova, V.,** *J. Electroanal. Chem.,* 129, 1, 1981.

140. **Roston, D. A., Shoup, R. E., and Kissinger, P. T.,** *Anal. Chem.,* 54, 1417AC, 1982.

141. **Kissinger, P. T.,** *J. Chrom.,* 488, 31, 1989.

142. **Horvai, G. and Pungor, E.,** *Crit. Rev. Anal. Chem.,* 21, 1, 1989.

143. **Stulik, K. and Pacakova, V.,** *J. Chrom.,* 208, 239, 1981.

144. **Stulik, K. and Pacakova, V.,** *Crit. Rev. Anal. Chem.,* 14, 267, 1984.

145. **Jandik, R., Haddad, P. R., and Sturrock, P. E.,** *Crit. Rev. Anal. Chem.,* 20, 1, 1984.

146. **Scott, R. P. W.,** *Liquid Chromatographic Detectors,* Elsevier, Amsterdam, 1986.

147. **Shoup, R. A.,** in *High Performance Liquid Chromatography,* Vol. 4, Horvath, C., Ed., Academic Press, New York, 1986, 91.

148. **Radzik, R. M. and Lunte, S. M.,** *Crit. Rev. Anal. Chem.,* 20, 317, 1989.

149. **Jones, P. G.,** *Anal. Chem.,* 57, 1057A, 1985.

150. **Yeung, E. S., Ed.,** *Detectors for Liquid Chromatography,* John Wiley & Sons, New York, 1986.

151. **Alfedredson, T. and Sheehan, T.,** *J. Chrom. Sci.,* 24, 473, 1986.

152. **Berry, V.,** *Crit. Rev. Anal Chem.,* 21, 115, 1989.

153. **Dose, E. V. and Guiochon, G.,** *Anal. Chem.,* 61, 2571, 1989.

154. **Fell, A. and Clark, B. I.,** in *Selective Sample Handling and Detection in HPLC,* (J. Chrom. Library Vol. 39A), Frei, R. W. and Zech, K., Eds. Elsevier, Amsterdam, 1988, 289.

155. **Orsulak, P. I., Sink, M., and Weed, I.,** *Ther. Drug. Monit.,* 6, 444, 1988.

156. **Hullet, F. I., Levy, A. B., and Tachiki, K. H.,** *J. Clin. Psychiatry,* 43, 165, 1982.

157. **Siebers, R. W., Chen, C. T., Terguson, R. I., and Maling, T. I. B.,** *Ther. Drug. Monit.,* 10, 349, 1988.

158. **Makino, T. A.,** *Clin. Chem.,* 29, 1313, 1983.

159. **Hogestam, I. H. and Pinkerton, T. C.,** *Anal. Chem.,* 57, 1757, 1985.

160. **Thomson, I. N.,** *Trace Anal.,* 2, 1, 1982.

161. **Sporn, M. B., Roberts, A. B., and Goodman, D. S., Eds.,** *The Retinoids,* Vol. 1, Academic Press, New York, 1984.

162. **Ballag, W.,** in *Retinoids, New Trend in Research and Therapy,* Saurat, I. H., Ed., S. Karger, Basel, 1985.

163. **Leenheer, A. H., Nelis, H. I., Lambert, W. E., and Bauwens, R. M.,** *J. Chrom.,* 429, 3, 1988.

164. **Wyss, R.,** *J. Chrom.,* 531, 481, 1990.

165. **American Chemical Society:** *Anal. Chem.,* 52, 2242 (1980).

166. **Fresh Fruit and Vegetables — Sampling: ISO 874,** International Standards Organisation, Geneva, 1980.

167. **Meat and Meat Products, Part 1 and 2, ISO 3100/1 and 2** International Standards Organisation, Geneva, 19705, 1980.

168. **Oilseeds — Sampling, ISO 542,** Oilseeds — Reduction of Laboratory Sample to Test Sample, ISO, 642, ISO, 1990.

169. **Pirkle, W. H. and Pochopsky, T. C.,** *Chem. Rev.,* 89, 347, 1989.

170. **Allenmark, S. G.,** *Chromatographic Enantioseparations; Methods and Applications,* Ellis Horwood, Chichester, England, 1988.

171. **Krstulovic, A. M., Ed.,** *Chiral Separations by HPLC: Applications to Pharmaceutical Compounds,* Ellis Horwood, Chichester, England, 1989.

172. **Lough, W. J., Ed.,** *Chiral Liquid Chromatography,* Blackie & Sons, Glasgow, 1989.

173. **Macaudiére, P., Lienne, M., Tambuté, A., and Caude, M.,** Pirkle-type and related chiral stationary phases for enantiomeric resolutions, in *Chiral Separations by HPLC: Applications to Pharmaceutical Compounds,* Ellis Horwood, Chichester, England, 1989, chap. 14.

174. **Lawrence, I. F. and Frei, R. W.,** *Chemical Derivatisation in Liquid Chromatography,* Elsevier, Amsterdam, 1976.

175. **Blau, K. and King, G. S.,** *Handbook of Derivatives of Chromatography,* Heyden and Son, London, 1977.

176. **Knapp, D. R.,** *Handbook of Analytical Derivatization Reagents,* John Wiley & Sons, New York, 1979.

177. **Imai, K. and Toyo'oka, T.,** in *Selective Sample Handling and Detection in HPLC,* Frei, R. W. and Zech, K., Eds., Elsevier, Amsterdam, 1988, 209.

178. **Frei, R. W., Jansen, H., and Brinkman, V. A.,** *Anal. Chem.,* 57, 1529A, 1985.

179. **Krull, I. S., Ed.,** *Post-Column Reaction Detection in HPLC,* Marcel Dekker, New York, 1986.

180. **Weinberger, R. and Femia, R. A.,** in *Selective Sample Handling and Detection in HPLC, Part A,* Frei, R. W. and Zech, K., Eds., Elsevier, New York, 1988, 395.

181. **Borch, R. F.,** *Anal. Chem.,* 47, 2437, 1975.

182. **Miller, J. M., Brindle, I. D., Cater, S. R., and So, K. H.,** *Anal. Chem.,* 52, 2430, 1980.

183. **Durst, H. D., Milano, M., Kitka, E. J., Connelly, S. A., and Grushka, E.,** *Anal. Chem.,* 47, 1797, 1975.

184. **Menasti, E., Gennaro, M. C., Sorzanini, C., Baiocchi, C., and Savigliano, M.,** *J. Chrom.,* 322, 177, 1985.

185. **Szepesi, G.,** *How to Use Reverse Phase Chromatography,* VCH, New York, 1992.

186. **ACS Committee on Environmental Improvement,** *Anal. Chem.,* 52, 2242, 1980.

187. **Peng, G. W. and Chiou, W. L.,** *J. Chromatogr.,* 531, 3, 1990.

Part VI

Radioanalytical Techniques

Chapter 21

Radioanalysis

Principle of the Technique

In radioanalytical techniques, radiation produced by unstable atomic nuclei, i.e., by radioactive isotopes, is made use of. Unstable atomic nuclei become stable ones with the emission of α, β, or γ radiation. Radioanalytical techniques are based on the measurement of the intensity and energy of this radiation. The unit of radioactivity is becquerel; its symbol is Bq. This is the activity of a source in which one disintegration occurs in a second. Accordingly, 1 Bq = 1 s^{-1}. The unit of radioactivity used earlier is the Curie (Ci), which is the activity of 1 g of radium; 1 Ci = 3.7×10^{10} Bq. Radioactive concentration is also used which is the activity of unit volume of liquid or gas.

The activity is proportional to the number of active nuclei:

$$A = \lambda N^*$$

where λ is the decay constant.

The change in the number of active nuclei with time is described by the following differential equation:

$$dN^* = -\lambda N^* dt$$

If at the starting time t = 0 the number of active nuclei is N_0, at a time t we have

$$N_t^* = N_0^* e^{-\lambda t}$$

i.e., the number of nuclei decreases according to an exponential function.

In practice, radioactive isotopes are characterized by the half-life of the radionuclide, i.e., the time required by the number of nuclei to decrease to half of their original number:

$$t_{1/2} = \frac{\ln 2}{\lambda}$$

In radioanalytical measurements frequently it is sufficient to apply the reproducibly determined pulse number or count number.

Measurement of the radiation is carried out on the basis of the interaction of radiation and matter. Upon the action of the radiation on the detector, a signal is produced whose number in unit time and whose magnitude are proportional to the intensity and the energy, respectively, of the radiation.

The types of detectors used most frequently are the following:

1. Ionization detectors. These enable the measurement of the intensity of highly ionizing α and β radiations, and — at a lower efficiency — also of γ radiations.

2. Scintillation detectors. NaI single crystals activated with Tl are most frequently used. Upon the action of highly penetrating γ- or X-rays, the atoms present in the scintillator get into excited state, and — according to the different energy transfer mechanisms — photo or Compton effect or pair formation occurs. The excited nucleus is stabilized with the emission of a photon. An electron multiplier (photomultiplier) coupled to the scintillator transforms and amplifies the light signals. The amplified pulses are processed in various measuring instruments. These detector types also enable the measurement of the energy distribution of γ-radiation. Their energy resolution is about 8%, referred to as the 0.66 MeV photo peak of Cs-137. Their efficiency is about 10%.

3. Semiconductor detectors. On the basis of their operation these can be regarded as solid-state ionization chambers. Upon the action of radiation, the resistance of the semiconductor is changed, and this is proportional to the magnitude of the energy transformed. Germanium or silicon-based semiconductor detectors, doped with lithium atoms, are frequently used. The impurity atoms are brought into the crystal lettice. This process is termed drifting. Ge(Li) or Si(Li) detectors of 50- to 100-cm^3 volumes can be prepared by this procedure. These days high purity germanium detectors (HPGe) are more widely used. The efficiency of these detectors is between 10 and 100 relative% [referred to as $3'' \times 3''$ NaI(Tl) scintillation detector] depending on the geometry. It is their advantage that their resolving power is 1.8 to 2.5 keV; however, their energy utilization efficiency is lower than that of scintillation detectors.

Processing the Signals of the Detectors

The signal supplied by the photomultiplier tube is conducted to the preamplifier built into the detector; signal transformation occurs at this stage and impedance matching to the main amplifier, the latter on account of the capacitive coaxial shielded cable.

If it is the frequency of particles sensed by the detector that is to be determined, a pulse counter or scaler is used. The basic unit of scalers is the bistable multivibrator which produces an output signal upon every second input signal and which consequently does binary counting. By connecting multivibrator circuits in series, scalers of any optional capacity can be produced; by application of passive circuit elements and diodes these can be converted to decadic counters. In the instruments, display of the stored pulses is realized by numerical display tubes which have ten stable states.

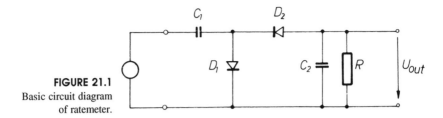

FIGURE 21.1
Basic circuit diagram
of ratemeter.

Frequently it is not necessary to record each pulse individually but it is sufficient to determine the mean number of pulses for a given time interval.

This can be carried out by rate meters. The input pulses are formed with respect to amplitude and duration, and hereupon conducted to an integrating circuit (Figure 21.1). The output signal of the latter (U_{out}) is proportional to the mean pulse number.

The measurement of the amplitudes, corresponding to different E values, independently of one another, or removal of interfering background radiation can be realized with energy-selective counters, the so-called single-channel spectrometers. The signals are sorted according to size and their frequency in time is measured. This task can be solved by means of amplitude discriminators.

The integral discriminator can easily be realized by a biased diode (Figure 21.2). These enable limiting the signals in one direction. It is apparent from the characteristic curve that the diode passes pulses of different heights, depending on the tension U_k applied across it.

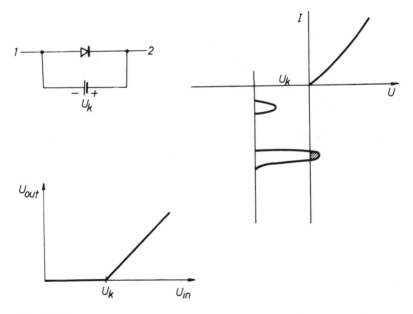

FIGURE 21.2
Basic circuit diagram and characteristic curve of integral discriminator. 1: Input, 2: output.

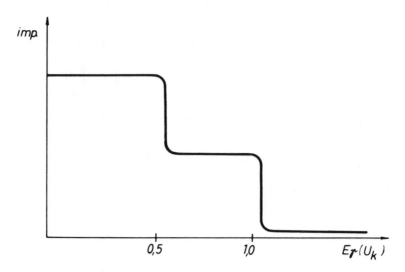

FIGURE 21.3
Integral gamma spectrum (E = 0.66 and 1.12).

If the height of the input pulse does not reach the value of U_k, the diode remains unconductive. On appearance of a pulse higher than U_k, the diode becomes conductive and such pulses are passed through it without further selection. By gradually increasing the value of U_k, the integral spectrum is obtained (Figure 21.3).

By application of a differential discriminator it can be attained that only pulses of predetermined height are measured. This can realized by two biased diodes connected with an anticoincidence unit (Figure 21.4). Only pulses of a height within the range ΔU are passed by this circuit. The instrument is designed in such a manner that the full energy range can be scanned by altering U_k at a given ΔU (channel width) value. In this manner, a differential spectrum is obtained (Figure 21.5).

By suitable adjustment of the values U_k and ΔU it is possible to measure the signals of the detector only within a given range and, in the case of a multicomponent system, to record pulses pertaining to one of the components.

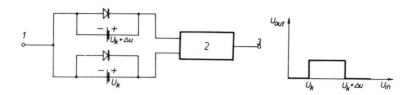

FIGURE 21.4
Basic circuit diagram and characteristic curve of differential discriminator.
1: Input, 2: anticoincidence, 3: output.

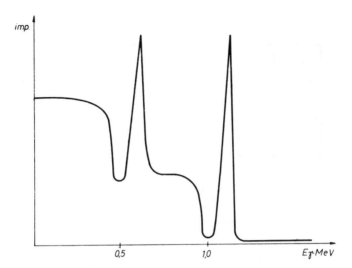

FIGURE 21.5

Differential gamma spectrum (E = 0.66 and 1.12).

In nuclear measuring techniques, multichannel analyzers, built up of digital circuits, are indispensable; these enable rapid determination of the distribution of the radiation with respect to energy. Essentially, the apparatus transforms the signals supplied by the detector, whose amplitude is proportional to the energy of the incident beam into pulses, the number of the latter being proportional to the signal amplitude (analog-digital converter). This is attained in such a manner that a capacitor is charged by the incoming signal through a diode up to a maximum value and hereupon the capacitor is discharged through a constant-current source. Simultaneously a gate circuit is opened so as to pass the pulses of a constant-frequency (100 to 425 MHz) clock generator. If the voltage across the capacitor reaches zero level, the gate circuit is closed and no more pulses are passed. The number of pulses passed is proportional to the magnitude of the amplitude to be measured, i.e., to the energy of the radiation. One pulse is now stored into the memory at a location (address or channel) corresponding to the number of pulses. The memory storing the impulses which is situated in the central unit of traditional analyzers is a card in the microcomputer. Data are processed by computer. The usual number of channels is 16 k, 8 k being used for data acquisition, and 8 k for data processing. The number of pulses stored in a given channel is proportional to the frequency of the given amplitude, i.e., to the intensity of the radiation measured (Figure 21.6). These processes go on following one another, but within a time of the order of microseconds and consequently the dead-time of multichannel analyzers is considerably longer than that of single-channel instruments. Multichannel analyzers can also be applied as multiscalers and enable continuous following of processes changing in time.

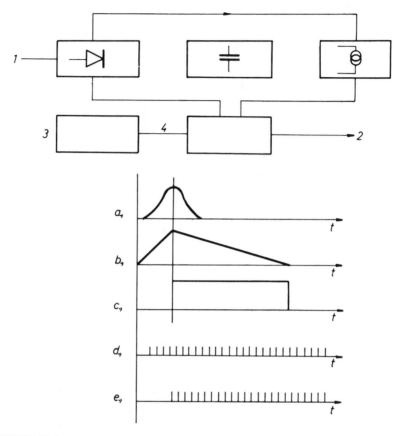

FIGURE 21.6
Basic circuit diagram and theoretical operation of an analog-digital converter.
1: Input, 2: output, 3: clock generator, 4: gate, a: shape of signal, charging capacitor through diode, b: process of charging and discharging capacitor, c: gate circuit, d: clock generator, e: gated pulses (address register).

Radioindication Techniques

The procedures of radioanalytical chemistry are based on the radioindication principle. The technique of radioindication or tracing is based on the fact that isotopes and radioisotopes of an element behave identically from a chemical point of view and that radioactivity can be detected by highly sensitive methods.

A tracer may be a chemical or a physical one depending on whether it is built into the molecule or else it is a mixture of isotopes that is used. The tracer isotope should always be in a chemical form identical to that of the element to be traced.

Tracer reagents can be applied in almost every one of the classical analytical techniques: radiogravimetry, radiochromatography, etc., but based on this a special radioanalytical method; the isotope dilution analysis has been developed.

The selectivity and sensitivity of tracer analytical techniques is in a number of cases considerably preferable to those of the classical method.

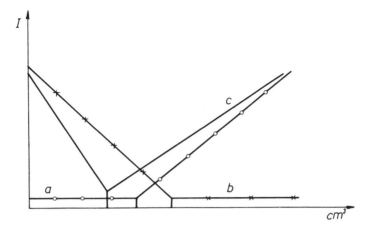

FIGURE 21.7
Precipitation-type radiometric titration curves.
I: intensity, cm³: volume of titrant, b: titrand labeled, c: both labeled, a: titrant labeled.

In *radiometric titrations* either the components taking part in the reaction or the indicator may be labeled. It is a requirement of basic importance that the reaction product or the indicator should enter another phase in the course of the titration, since the total activity of the solution remains unchanged. This occurs per se in the case of precipitation titrations (Figure 21.7). In complex formation titrations, the reaction product is frequently separated by extraction or ion exchange.

Titration of multicomponent systems can also be carried out if the stability constants or extraction constants of the individual components are sufficiently different. For example, mercurous, silver, and zinc ions can be titrated with dithizone if the solution is labeled with Hg-203 and Zn-65 radioactive tracers and the changes in the specific activity of the aqueous or of the titrant added (Figure 21.8). This technique enables a few micrograms of a metal ion to be determined with a relative error of ±1%.

The basic principle of *isotope dilution techniques* is the following: if an inactive isotope is added to a radioisotope, the activity of the mixture remains unchanged whereas the specific activity, as referred to the species in question, decreases. The amount of the active or of the inactive species can be calculated from this change in activity. It follows from the principle of the technique that a fraction of the "diluted" substance is to be separated and the specific activity or intensity of this fraction of known mass (or volume) is to be measured in the course of the determination. Any known chemical procedure can be used for separation: at the same time, the latter necessarily determines the accuracy and sensitivity of the measurements. The isotope dilution technique is generally not applicable as a microanalytical technique: its main advantage is rapidity.

Simple isotope dilution: the own radioactive isotope is added to the element or compound to be determined in a known amount, knowing the specific activity.

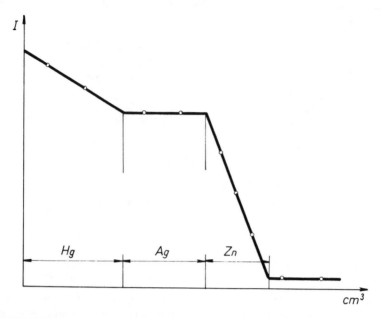

FIGURE 21.8

Complex formation-type radiometric titration. 1: intensity of aqueous phase, cm³: volume of dithizone solution (titration of Hg^{2+}, Ag^+, and Zn^{2+} with a chloroform solution of dithizone) titrand labeled with Hg-203 and Zn-65.

If the amount of the unknown substance is x and y g of its radioactive isotope is added to it, whose intensity is I_0, and z g of the mixture is precipitated and the intensity of the precipitate is I_1, the following equations may be written:

$$(x + y) : I_0 = z : I_1$$

$$x = z\frac{I_0}{I_1} - y$$

If $y \ll x$, i.e., the mass of the radioactive isotope is negligible as compared to that of the component to be determined (in the case of application of a carrierless tracer), we may write

$$X = z\frac{I_0}{I_1}$$

Expressed with specific activities, when in the case of the radioactive isotope

$$S_0 = I_0/y$$

and after dilution

$$S = I/z$$

wherefrom

$$x = y(S_0/S_1 - 1)$$

Reverse isotope dilution — The amount or the radioactive substance together with its carrier is determined in such a manner that it is diluted with its inactive isotope.

Any g amount of inactive substance is added to the radioactive substance to be determined, the amount of the latter being x g and its intensity I_0 (its specific intensity $S_0 = I_0/x$). The intensity of a quantity z, after separation and mixing is I_1; hence $S_1 = I_1/z$ and

$$(x + y):I_0 = z:I_1$$

$$x = \frac{y}{S_0/S_1 - 1}$$

Double isotope dilution — This solution is used when the amount of radio element is extremely small which hinders the separation. Two different amounts of the inactive isotope are added to two equal portions of the radioactive isotope. The active-inactive isotope ratio will be different after separation, i.e., different specific intensities will be observed and hence the concentration of the radioactive substance can be calculated.

y_1 and y_2 g portions of the inactive isotope are added to two identical portions of the x g radioactive isotope to be determined. After separation, the specific intensities are measured:

$$S = I_1/z_1 \quad \text{and} \quad S_2 = I_2/z$$

$$x = \frac{y_2 S_2 - y_1 S_1}{S_1 - S_2}$$

Substoichiometric isotope dilution — This is the most important of the isotope dilution techniques which can be regarded as the chemical method of the highest sensitivity. It is a variant of isotope dilution in which an amount lower than the stoichiometric one of the reagents used for separation is added to the component to be determined. This ensures that an identical amount of substance is always separated from the solutions of different concentrations.

If a reagent (e.g., dithizone in carbon tetrachloride) is added in a substoichiometric amount to a radioactive solution containing zinc in a known amount (y) and the intensity (I_0) of the solution containing the reaction product (the organic phase) is determined after the separation, and hereupon the solution containing the unknown amount of zinc to be determined is added to the solution containing the above-mentioned amount of radioactive zinc, an identical amount of reagent is added and the intensity of organic phase (I_1) is measured under identical circumstances after separation, the following connections may be written:

$$I_0 : I_1 = (x + y) : y$$

$$x = y(I_0/I_1 - 1)$$

The sensitivity of this technique may in some cases be as high as 10^{-9} g/cm^3, e.g., in the determination of Zn, Hg, Fe, and Cu. Its advantage is the low time requirement of the measurement and its high selectivity.

The field of application is the determination of trace inorganic components.

Activation Analysis

The basic principle of the technique is that stable isotopes are transformed to radioactive isotopes upon the action of radiations initiating nuclear reactions and qualitative and quantitative analysis of the products can be carried out the basis of their characteristic radiation.

It is neutrons that bring about nuclear reactions at the highest probability because they are uncharged particles. Neutrons especially at the thermal energy level (the most probable energy 0.025 eV) bring about reactions which are preferable from the chemical point of view, since atomic nuclei more abundant in neutrons are produced with simultaneous γ radiation. Such nuclei are unstable and they get stabilized with the emission of radioactive radiation.

This is the neutron-gamma (n,γ) nuclear reaction:

$$^A_Z X (n, \gamma) \; ^{A+1}_Z X$$

Fast neutrons possessing larger energy cause the nuclear reactions of fast neutrons: (n,p) and (n,α). However, these occur at a much lower probability than the (n,γ) reaction, i.e., the σ activation cross section is much smaller.

$$^A_Z X (n, p) \; ^A_{Z-1} Y$$

$$^A_Z X (n, \alpha) \; ^{A-3}_{Z-2} Y$$

In some cases, these nuclear reactions may be the basis of a determination, e.g., ^{58}Ni (n,p) ^{58}Co, but generally they should be regarded as interfering reactions. The isotopes formed in the nuclear reaction decay at a rate characteristic of their quality and a new radioactive isotope or a stable nucleus is produced.

The rate of the production of radioactive nuclei depends on the number of target nuclei present in the sample (N), the activation cross section of the nuclear reaction (σ), and the neutron flux (Φ) of the radiation source.

In parallel with the production of active nuclei, their decay also starts and accordingly the rate of accumulation of active nuclei is determined by the difference in the rates of formation and decomposition:

$$\frac{dN^*}{dt} = \Phi \sigma N - \lambda N^*$$

where N is the number of target nuclei:

$$N = \frac{m N_A f}{M_A}$$

where Φ is neutron/$m^2 s^1$, m is the mass of the target, N_A is 6×10^{23} atoms/g atom, f is the weight fraction of the isotope occurrence, M_A is the mass number of the target nucleus, λ is the decomposition constant, and σ is the probability of the occurrence of the nuclear reaction, its unit being 1 bar = 10^{-28} m^2.

At the end of irradiation, the number of active nuclei is

$$N_{t_b} = \frac{\Phi \sigma N}{\lambda} (1 - e^{-\lambda t_i})$$

where t_i is the duration of irradiation

The activity is expressed by the number of nuclei decomposed per second:

$$A_{t_b} = N_{t_b}^* \lambda = \Phi \sigma N (1 - e^{-\lambda t_i})$$

where $1 - e^{-\lambda t_i} = S$ (saturation factor)

After a time t_c the activity is

$$A_{t_c} = \Phi \sigma N S e^{-\lambda t_c}$$

where $e^{-\lambda t_c} = D$ (decay factor)

If the measurement time is commensurate with the half-life time of the isotope measured, the decomposition of the isotope during the measurement time is to be taken into account, i.e., the measured activity which is an integral average is smaller than that at the beginning; hence, the average activity during the measurement time t_m will be

$$A_{ave} = \Phi \sigma N S D \frac{1 - e^{-\lambda t_m}}{\lambda t_m}$$

where

$$\frac{1 - e^{-\lambda t_m}}{\lambda t_m} = C_m \text{ (counting factor)}$$

The temporal changes in activity during activation, decomposition, and measurement are shown in Figure 21.9.

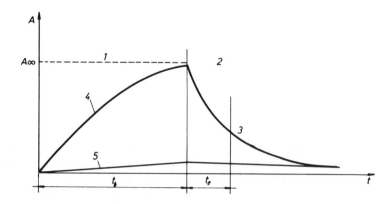

FIGURE 21.9
Changes in activity depending on activation and decomposition time.
1: Activation, 2: decomposition, 3: measurement, 4: short $t_{1/2}$, 5: long $t_{1/2}$.

The average activity can be calculated from the peak area measured as follows:

$$A_{ave} = \frac{n}{t_m \, \varepsilon}$$

where n is number of counts in the full energy peak and ε is efficiency of the full energy peak.

From these the mass to be determined will be

$$m = n \frac{M_A}{f N_A \, \sigma \varepsilon} \; \frac{1}{\Phi} \; \frac{1}{t_m \, SDC_m}$$

Absolute, Relative, and Comparator Methods

In the absolute method all the quantities on the right-hand side of the equation used for calculating the mass of the element are measured while the constants are taken from tables and the mass is calculated. An obvious advantage of the method is that there is no need for a reference material and the result is based on a single measurement. The drawback is that the actual neutron flux and the absolute efficiency have to be measured. The latter quantities can be measured; however, great care has to be taken and the errors of these measurements are included in the final results. In addition, part of the activation cross-section data taken from tables has a significant error which leads to an appreciable systematic error in the results. This is the reason why this method is used less frequently than the other methods.

   In the relative and comparator methods a standard material is irradiated together with the sample under the same experimental conditions and the mass of the component to be determined is obtained by comparing the experimental data of the two materials.

The relationship for the calculation of the unknown mass can be derived as follows: the mass of the standard component (m*) is expressed by the above equation denoting all the specific parameters by * and the equation for the unknown component is divided by that for the standard component. The unknown mass can be expressed as follows:

$$m = m* \ \frac{n}{n*} \ \frac{f* \sigma* \epsilon*/M_A^*}{f \sigma \epsilon/M_A} \ \frac{\Phi*}{\Phi} \ \frac{t_m^* \ S* D* C_m^*}{t_m \ S D C_m}$$

$\Phi/\Phi*$ is the ratio of the neutron fluxes which is equal to 1 as the irradiation conditions are to be the same.

By rearranging this equation we get

$$\frac{f \sigma \epsilon/M_A}{f* \sigma* \epsilon*/M_A^*} = \frac{n}{t_m \ S D C_m \ m} \ \frac{t_m^* \ S* D* C_m^* \ m*}{n*} = \frac{i_{sp}}{i_{sp}^*} = k$$

The basic difference between the relative and comparator methods is in the nature of the element used as the standard. In the relative method the standard element is the same as the one to be determined, in the comparator method a different element is used. Using the analytical terminology the relative method may be called single-point calibration, the comparator method internal standard method.

It follows from what has been given above that for the relative method k = 1 (provided the measurement geometry is the same; otherwise differences are to be corrected for), for the comparator method k <> 1. Accordingly (and considering that $\Phi*/\Phi = 1$) the relationship for the calculation of the unknown mass will be

$$m = m \ \frac{n}{n*} \ \frac{t_m^* \ S* D* C_m^*}{t_m \ S D C_m}$$

In using the comparator technique a known amount of the element to be measured is activated together with a known amount of the comparator element under identical conditions, and k is calculated from the experimental data. If the analytical measurement and the determination of k are carried out in different irradiation channels (possibly at different thermal/epithermal flux ratios), further measurements and calculations are needed.

The relative and the comparator methods have approximately the same accuracy and reproducibility. Selection is made according to the task. In general the relative method is chosen if a well-defined and reliable standard material is available and the number of elements to be determined is relatively small. The use of the comparator method is justified if a large number of multicomponent samples are to be analyzed due to the painstaking work required for the precise determination of k values.

Activation analysis enables to attain a high sensitivity (10^{-6} to 10^{-12} g) and this parameter is largely dependent on the flux of the neutron source used for irradiation. The most abundant neutron source is an atomic reactor in which

$$\Phi_{therm} \sim 10^{16} \text{ to } 10^{20} \text{ m}^{-2}\text{ s}^{-1}$$

$$\Phi_{fast} \sim 10^{14} \text{ to } 10^{15} \text{ m}^{-2}\text{ s}^{-1}$$

The effect of thermal neutrons can be filtered off in the course of irradiation if the sample is wrapped in a cadmium foil or the irradiation vessel is embedded in boron powder. This is necessary in order to separate various nuclear reactions.

The Course of Activation Analysis

Prior to irradiation, calculations are made — on the basis of the known data of the sample — on the expected radiation level (this is also necessary in connection with radiation hazards).

The optimum parameters of irradiation are approximately determined on the basis of the nuclear data of the components are introduced into the sample in the course of any chemical operations prior to irradiation. Dissolution or digestion of the sample should, if possible, be carried out after activation, having added a larger amount of the inactive atoms of the component to be determined (carrier technique). In this case, impurities present in the solvents and reagents do not interfere with the determination: this is a great advantage of activation analysis as compared to other trace analytical techniques.

Two variants of the activation analysis technique are used in practice.

If in consequence of the nuclear properties of the radioisotopes produced in the sample can be measured directly and in a selective manner with an instrumental method (γ-spectrometry), the nondestructive procedure is applied. If such is not the case since, for example, the γ-photon energies of the components are nearly identical or the difference in relative intensities is high, e.g., the main components of the sample also get activated to a high degree (e.g., metal matrices or biological materials), it is necessary to separate low-activity components present at trace levels.

In such cases, since radiation measurement is preceded by chemical separation, the technique is termed a destructive activation analytical one.

In the case of the destructive technique, the aim is the individual or group separation of the elements or radioisotopes, corresponding to a degree to the selectivity and sensitivity of nuclear measuring technique.

Determination of the individual components of the activated sample is carried out by the measurement of the energy and intensity of the radiation emitted by the radioisotopes.

In order to make qualitative identification, the energy distribution of the γ radiation is measured. This can be carried out by the γ-spectrometric technique, with the aid of multichannel analyzers. The γ-spectra are recorded with the samples prepared for measurement with a detector of adequate and placed in a location of a background radiation as low as possible. Having carried out qualitative identification (determination of the E values pertaining to the full-energy peaks observed and identification of the corresponding radioisotope).

In the lower energy ranges, the full-energy peaks appear superimposed upon Compton and other effects and this makes quantitative evaluation difficult or even impossible.

A full-energy peak, if there is no interference, follows the Gaussian distribution: the peak area or the sum of the pulses stored away in the channels is proportional to the amount of the radioisotope producing the peak. The net peak area is generally calculated by the TPA (total peak area) method in which the background is approximated by a trapese.

As far as the analytical application of activation analysis is concerned the following statements can be made.

The detection limit of neutron activation analysis is very good (down to the ppm, ppb level) for many elements but not for all of them. A very important feature of the method is that it is theoretically well established. These features ensure that NAA is a powerful analytical method and, among a variety of other applications, it is one of the most important methods in the certification of reference materials, and plays a very important role in the solution of special problems (analysis of high purity materials, rare earth elements, etc.). However, the technique requires special apparatus and conditions; hence, its use is limited. Its use is also hindered by the recent tendencies to avoid methods based on radioactivity.

21.1. Determination of Zinc with the Substoichiometric Isotope Dilution Technique

Principle of the Determination

In the case of substoichiometric isotope dilution both the solution containing the radioactive tracer and its solution diluted with its inactive isotope are made to react with identical and substoichiometric amounts of the reagent and the intensities of equal portions of the reaction products are measured. In this case it is not necessary to calculate with specific activities since the following equation may be written for the intensity rations:

$$I_0 : I_1 = (x + y) : y$$

and

$$x = y(I_0/I_1 - 1)$$

where x is the mass of Zn in the added volume of the solution of unknown concentration (μg), y is the mass of Zn in the added volume of the radioactive diluted stock solution of known concentration containing the tracer (μg), I_0 is the intensity of the separated organic phase prior to dilution (s^{-1}), and I_1 is the intensity of the separated organic phase after dilution (s^{-1}).

Apparatus

Energy selective scaler with scintillation detector
Separating funnels 100 cm^3
Graduated test tubes
Pipettes, 2, 5, and 10 cm^3

Chemicals

Zinc sulfate stock solution labeled with Zn-65 radioisotope of known concentration, whose concentration is 100 μg Zn/cm^3. The solution is prepared of a $^{65}ZnSO_4$ preparate with deionized water, from which all traces of Zn were removed with dithizone, by extraction. Prior to application, its Zn content is accurately determined (either by titration with dithizone or else by inverse isotope dilution technique). A 10-μg/cm^3 solution is prepared of the stock solution with water that was shaken with dithizone. The solution is stored in a plastic bottle; 10% sodium acetate solution, shaken with dithizone; and 0.001 M dithizone in chloroform are used.

Aqueous or slightly acidic solution of zinc sulfate to be determined; concentration 2 to 5 μg Zn/cm^3.

Procedure

10 cm^3 deionized water, purified with dithizone, is poured into a 100-cm^3 separating funnel and 2.0 cm^3 of the diluted stock solution of known concentration (20 μg Zn) labeled with Zn-65 is added. Hereupon 3 cm^3 sodium acetate solution, 2 cm^3 dithizone solution in chloroform, and 8 cm^3 chloroform are added, vigorously shaken and the two phases are allowed to separate. 5 cm^3 of the organic phase is transferred into a graduated test tube and the latter is placed into a detector equipped with a well-type scintillator. The intensity (I_0) is determined.

A 10.00-cm^3 portion of the solution containing the Zn is placed into another separating funnel, and — except for the water — the solutions mentioned in the foregoing are added and the contents are shaken. 5 cm^3 of the organic phase is brought into a graduated test tube and the intensity (I_1) is measured.

The Zn content of the unknown Zn solution is calculated with the formula given in the above. (y = the Zn content of 2 cm^3 of the labeled diluted Zn stock solution in μg.)

When measuring the intensity, the amplification of the energy selective scaler is adjusted in such a manner that the photopeak of Zn-65 (E) = 1.119 MeV. The measurement time is adjusted so as to obtain about 10,000 counts after background correction; in this manner the error in the radiation measurement is about 1%.

Evaluation

Measure the background radiation with and without suppression with the lower threshold. Calculate the error in the determination of I_0 and I_1.

Calculate the Zn content of the unknown solution in $\mu g/cm^3$ units.

21.2. Analysis of Steel by Neutron Activation Analysis

Principle of the Measurement

On irradiating a steel sample containing manganese and vanadium by thermal neutrons, (n,γ) reactions take place and the nuclear reactions result in radioactive isotopes. The isotopes originally present and those formed in the nuclear reaction as well as their parameters are presented in the following table:

Target nucleus	^{55}Mn	^{51}V	^{58}Fe	^{54}Fe
Radioactive isotope produced	^{56}Mn	^{52}V	^{59}Fe	^{55}Fe
Abundance (%)	100	99.75	0.31	5.84
Activation cross section 10^{-28} m²	13.3	4.8	1.23	2.8
Half-life time	2.58 hours	3.76 m	45.1 d	2.6 y
γ-energy (intensity) (keV)	846.9 (100) 1810.7 (25) 2112.8 (15)	1434.1 (100)	1098.6 (100) 1291.5 (80)	

The most important difference between the nuclear parameters of the radionuclides produced appears in the half-life time. With respect to the degree of activation, the abundance of the target nuclei in the natural isotope mixture and the activation cross section are also of importance. Choosing a short irradiation time — in the order of minutes — at a thermal neutron flux of 2.7×10^{15} m^{-2}s^{-1} ^{56}Mn and ^{52}V isotopes reach appreciable activities while the activity of ^{59}Fe (having a long half-life time compared with the irradiation time) which is produced of ^{58}Fe (abundance in natural iron only 0.31 %) at a relatively low activation cross section, remains below the detection limit. The significant difference between the half-life times of ^{52}V and ^{56}Mn and the fact that the

1434.4-keV peak of ^{52}V is superimposed on a low ^{56}Mn background allows the simultaneous instrumental determination of Mn and V without any mathematical manipulation. The γ-spectrum of the activated sample measured after a short cooling time enables the full-energy peak of ^{52}V to be evaluated while after a longer cooling period — about 40 min — the ^{52}V peak does not appear in the spectrum, hence the 846.9-keV peak of ^{56}Mn can be evaluated free of the interference of the Compton-region of ^{52}V. It should be noted that this procedure is needed mainly if the γ-spectrum is taken using a scintillation detector. When a semiconductor detector is used — since its resolution is about 50 times that of the scintillation detector — the ^{56}Mn peak superimposed on the Compton region of ^{52}V can be evaluated with appropriate accuracy.

Apparatus

Nuclear reactor with minimum 2×10^{15} m^{-2}s^{-1} thermal neutron flux
HPGe semiconductor detector
Amplitude analyzer, minimum 4 kByte
Printer
PC, or programmable calculator

Procedure

1. The expected activity is estimated from the sample mass, probable concentration, and irradiation time.

2. The energy calibration curve is measured using closed standard sources. The standards and their γ energies are as follows:

Isotope	^{137}Cs	^{60}Co	^{88}Y
γ-energy (keV)	661.6	1173.1	898.0
		1332.4	1836.1

3. The encapsulated steel sample is irradiated and the γ-spectrum recorded. The peaks are identified based on the peak energies using the energy calibration.

4. The V and Mn standards are irradiated in succession. The spectrum of V is taken while Mn is irradiated, and the spectrum of the Mn standard is taken immediately afterward.

5. The spectra are evaluated. The peaks are marked and peak areas calculated. The amount of V and Mn in the sample is calculated from the amounts of standards taking the irradiation, cooling, and measurement times into account. The concentrations are calculated based on the sample weight.

21.3. References

De Soete, D., Gijbels, R., and Hoste, J., *Neutron Activation Analysis,* Vol. 34, Elving, P. J. and Kolthoff, I. M., Eds., Wiley-Interscience, London, 1972.

Lieser, K. H., *Einführung in die Kernchemie in Einzeldarstellungen,* Verlag Chemie GmbH, 1969.

Ruzicka, J. and Stary, J., Eds., *Substoichiometry in Radiochemical Analysis,* Pergamon Press, London, 1968.

Tölgyessy, J., Braun, T., and Kyrs, M., *Isotope Dilution Analysis,* Akadémiai Kiadó, Budapest, 1972.

Part VII

Flow Analytical Techniques

Chapter **22**

Voltammetric Measurements in Streaming Solutions

Principle of the Techniques

The voltammetric measuring technique where the component to be reduced or oxidized reaches the electrode surface by convection is termed hydrodynamic voltammetry. The convection can be the result of the rotation or vibration of the working electrode, but the stirring or the streaming of the sample solution is also used frequently. Hydrodynamic voltammetry is a well-applicable technique for following different physical or chemical reactions. In certain cases the solution to be analyzed is a moving one, while in other cases the sample solution is streamed for the sake of continuous analysis. In addition to these, the hydrodynamic voltammetry has so many advantages that it is worthwhile to bring into stream the sample solution even in those cases when the sample solutions are individual entities.

It has to be noted that in flow-through analysis electrodes of constant surface area are most often used.

The advantages of voltammetric measurements under convective diffusion conditions are the following:

> Under convective diffusion conditions the current intensities — which are the quantity used for the concentration measurement — are higher, than under diffusion conditions.

> Since the measurements are generally performed with working electrodes of constant surface area the disturbing condenser current is smaller than using dropping mercury electrodes.

The hydrodynamic voltammetric measurements are carried out generally at constant working electrode potential, which lowers further the
value of the condenser current. Since the analysis is performed at
constant potential, the condenser current is more or less the same in
the course of measurement; thus, it can be compensated electronically.

The voltammetric limiting current intensity measured at a constant potential (i)
can be described (if the electrode reaction is reversible) as follows:

$$i = k\,n\,F\,D^{2/3}\,v^{-1/6}\,L\,c\,v^{z}$$

where k is constant, n is the number of electrons taking part in the electrode
reaction, F is the Faraday number, D is the diffusion coefficient of the
electroactive component, v is the kinematic viscosity of the streaming solution,
L is a constant characteristic to the size and geometry of the electrode, v is the
flow rate of the streaming solution, and z is a constant value characteristic to
the shape and position of the working electrode.

It is apparent from the formula that the limiting current intensity is strongly
dependent on the flow rate of the streaming solution.

22.1. Study of the Dissolution Rate of Drug from Pharmaceutical Preparations by Recycling

Principle

The dissolution rate of a drug from the pharmaceutical preparations is one of
the important characteristics. Most often a fast drug release is required, but in
the case of sustained release or retard preparations the slow dissolution rate is
the basis of their proper effect.

For dissolution rate measurements the Pharmacopoeas give detailed de-
scriptions. The official methods can be carried out in commercial equipments,
and they are used mainly in the quality control laboratories. The degrees of
automation of these equipments are different, and different analytical methods
are used for the determination of the concentration of samples taken from the
dissolution vessel from time to time, or continuously.

The aim of the experiment is to use a very simple dissolution-rate studying
system, and through this to get some experience with flow-through analytical
systems.

Apparatus

Thermostated dissolution vessel (1000 ml; see Figure 22.1)
Ultrathermostat with thermometer

Polarograph

Flow-through voltammetric cell (it can be a piece of tube with a built-in small graphite working electrode and a silver wire coated with silver chloride)

Peristaltic pump

Chemicals

Solution 0.1 mol/l in potassium chloride and 0.01 mol/l in hydrochloric acid (dissolution medium and at the same time the supporting electrolyte for the voltammetric measurement)

Chlorpromazine hydrochloride substance

Pharmaceutical preparation, e.g., chlorpromazine-containing tablet (25 mg)

Procedure

The apparatus shown in Figure 22.2 is assembled and 900 ml of the dissolution medium is poured into the dissolution vessel (1). The stirrer (2) peristaltic pump (4), and the polarograph attached to the measuring cell are started. It is checked whether all of the air bubbles have been removed from the sampling tube section (7) and measuring cell (5). If the temperature of the dissolution medium is 37°C, 25 mg of the chlorpromazine substance is introduced into the liquid and dissolved. Changing gradually the working electrode potential the appropriate potential is chosen. The achieved constant current value will show the 100% dissolution of the drug. Then the measurement is stopped. After rinsing the system and adjusting the measuring parameters chosen on the basis of the previous experiment, the dissolution medium is poured into the vessel, and when the temperature of the solution is 37°C a tablet to be examined is dropped into the solution. The current intensity is recorded in the course of the dissolution process. After dissolution of the total amount of the active material (the current intensity achieves a constant value) the experiment is stopped.

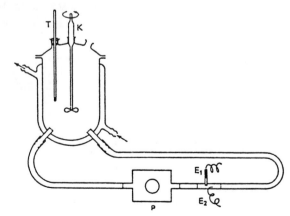

FIGURE 22.1
Thermostated dissolution vessel.

Evaluation

On the basis of the voltammetric current intensity vs. time curve determine

1. The time required for the dissolution of 90% of the drug from the preparation
2. The dissolved active material of the preparation.

In this experiment it is supposed that the relationship between the current intensity and the concentration of the drug is linear, and that the other ingredients of the preparation do not disturb the amperometric determination. Generally these assumptions can be considered as good approximations.

22.2. Study of the Dissolution Rate of Drug from Pharmaceutical Preparations by Flow Injection Analysis

Principle

The study of the dissolution rate by recycling is advantageous mainly in the case of fast dissolution pharmaceutical preparations. If the dissolution is slow, there is no need to control the drug concentration continuously. In such cases sufficient information can be obtained by taking samples from the dissolution medium from time to time, and analyzing them. For the analysis of series of samples a flow-through analytical technique, flow injection analysis can be used advantageously. (For details of the technique, see Chapter 23.)

Apparatus

The experimental set-up applicable for this kind of measurements is shown in Figure 22.2. The system can be built from the components described in Section 22.1. The construction of the system as well as the parameters (e.g., flow rate, length and volume of the mixing coil, the volume of the sample injected) can be chosen depending on the task to be solved. In this case an amperometric detector is used; however, others (e.g., UV photometric) can also be used.

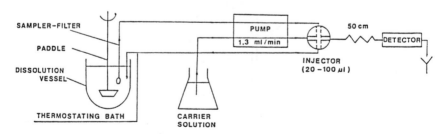

FIGURE 22.2
Experimental set-up for drug dissolution studies.

Chemicals

Solution 0.1 mol/l in KCl and 0.01 mol/l in HCl (dissolution medium and at the same time supporting electrolyte for the amperometric measurement)

Pharmaceutical preparation containing 500 mg N-acetyl-p-aminophenol 5×10^{-2} mol/l stock solution in N-acetyl-p-aminophenol

Set of calibration solutions (2×10^{-4}; 5×10^{-4}; 1×10^{-3}; 3×10^{-3}; 5×10^{-3} mol/l) from the stock solution by dilution with distilled water

Procedure

The system is assembled according to Figure 22.2, and 900 ml of dissolution medium is poured into the dissolution vessel. The stirrer, the peristaltic pump, and the measuring instrument (a polarograph in this case) are started. The potential — determined in a separate experiment — is adjusted. If the temperature of the dissolution medium stabilizes at 37°C, a tablet to be studied is dropped into the solution. The dissolution medium is circulated continuously through the injector. The loop of the injector is turned from time to time; thus a plug of sample is introduced into the carrier stream.

Evaluation

Read the peak height values. Prepare a calibration curve by plotting the peak heights vs. the concentration of the injected solutions.

Determine the concentration of the sample solutions on the basis of the calibration curve. Plot the drug dissolution curve by plotting the concentration of the drug in dissolution medium vs. the time. Prepare a calibration curve for the evaluation of the dissolution curve. For this purpose dilute the stock solution with distilled water by the application of the burette between the concentrations of 10^{-4} mol/l and 6×10^{-3} mol/l. Inject the solutions of different concentrations into the streaming supporting electrolyte. The height of the peaks vs. concentration of injected solution gives the calibration line.

22.3. References

Pungor, E., Fehér, Zs., Nagy, G., and Tóth, K., Automatic electrochemical analysis. II, *Crit. Rev. Anal. Chem.,* 14, 175, 1983.

Stulík, K. and Pacáková , V., *Electroanalytical Measurements in Flowing Liquids,* Ellis Horwood, Chichester, England, 1988.

Chapter 23

Automatic Laboratory Analyzers

Principle

The rapidly increasing demand of various fields of industry, agriculture, environmental protection and control, and medicine for analytical data led to the development of automatic analytical techniques. The technique enabled not only the time and cost of analysis to be reduced but also the sample throughput and reproducibility of the results to be increased.

Automatic analyzers are mainly used in fields where a high number of samples with similar matrices are routinely analyzed for several components, e.g., clinical and agricultural chemistry and environmental protection and control. The considerable costs of investment and method development are overcompensated by the reduction of the running costs due to the smaller reagent and energy consumption owing to the small amount of sample used, and the smaller amount of human labor, laboratory space, and glassware needed. Consequently, the cost per analysis is reduced.

Automatic laboratory analyzers belong to two main types:

> Batch-type or discrete-sample analyzers
> Continuous flow analyzers

In *batch-type analyzers* each sample is assigned a separate sample cup and treated as a separate entity through all steps of the analysis, dilution, reagent addition, mixing, heat treatment, centrifugation, etc. Sample cups are placed on a conveyor belt or turntable. At the end of the way each sample is presented in succession to the detector.

Batch analyzers simulate manual operations; thus, established manual methods are easily adapted to them. The advantage of these systems is that cross-contamination of samples is excluded, and the amount of reagents necessary is extremely small. The drawback is the mechanical complexity of the analyzer containing many moving parts which have limited lifetimes.

In *continuous flow analyzers* the sample becomes part of a stream of liquid and all the operations from sample dispensing to detection are carried out in streamed solution. In addition to operations like sample dispensing, dilution, reagent addition, and heat treatment which are equally well executed in batch analyzers, sample dialysis, distillation, solvent extraction, and other types of separation can also be carried out in the flowing system.

Continuous flow analyzers consist of a series of modules, each performing a specific task. Modules can be interchanged and rearranged to perform different analytical tasks, which makes this type of analyzer extremely versatile.

The basic part of a continuous flow analyzer is a peristaltic (proportionating) pump which propels solutions (samples, buffers, diluents, reagents, etc.) through flexible tubes. The constancy of the pump rate and of the tube diameters is a prerequisite of reliable and reproducible analytical results. The pump, depending on the type, may accommodate up to 28 channels. The number of channels used depends on the actual determination being made.

In the analytical unit samples are brought together with the reagent(s), and further operations are carried out to bring the analyte into a measurable form. At each point where two solutions come together a mixing coil of appropriate length is attached to allow mixing and reaction time to produce the detected species in a readily measurable concentration. In some cases (if a higher dispersion is required or the samples have high and varying densities and/or viscosities), a mixing chamber may be used.

At the end of the path the liquid flows through the detector which continuously measures some property depending on the concentration of the analyte. Various detection principles have been used; of electroanalytical techniques potentiometry and amperometry are most often used. Optical methods, flame photometry, atomic absorption, and absorption spectrophotometry are also employed, the last technique being by far the most widely used of all techniques.

The analytical signal is recorded by a strip-chart recorder or sent to a computer where it can be stored and accessed by a video terminal for surveying and by a printer for producing a report in the required form. The strip-chart recorder is the minimum requirement, but even if a computer is attached, it is very useful as it follows the curve produced by each sample and reflects any defect in the flow system which can then be alleviated.

Continuous-flow analyzers belong to two groups:

> Segmented-flow analyzers
> Flow-injection analyzers (FIA)

In a *segmented-flow analyzer* the sample is continuously aspirated for a predetermined time, then following a wash solution the next sample is aspirated. To prevent intermixing of subsequent samples, the sample stream is divided into several segments by air bubbles added in a regular manner. This allows fairly sharp leading and following edge of the sample plug to be maintained even along a long path. Before detection a debubbler is used to allow detection in a continuous stream of liquid. The sample can be divided into several parts to enable several determinations to be carried out from the same sample simultaneously, using as many analytical channels as there are components to be determined. This type of analyzer usually has a rather long start-up and shut-down time (0.5 to 1 hour). For these reasons this type is mainly used for analyzing large series of samples for several constituents, and when several steps (separations, reactions) are needed to transform the component to be determined into a sensitively and selectively detected form. The sample throughput may reach 120 samples per hour or higher, but it is typically 60 to 80 samples per hour.

In a *flow-injection analyzer* a plug of the sample 10 to 500 µl in volume is injected into a continuously streaming liquid (carrier solution). The sample forms a zone which is transported through the system to the detector which continuously detects a property of the solution which depends on the concentration of the analyte. The detector may detect the analyte in its original form (e.g., determination of hydrogen ions by a pH-sensitive glass electrode), or the product of a suitable reaction of the analyte.

The most important feature of flow injection technique is the controlled dispersion of the sample zone which allows highly reproducible results to be obtained.

A FIA system contains an injector in addition to the parts which all continuous flow analyzers contain, which may be manually operated or automatic. Usually one component of the sample is determined. This type of analyzer has a short start-up and shut-down time (a few minutes), so it is reasonable to use the system even if only a few samples are to be analyzed at a time. It is extremely simple to rearrange the system for the determination of another constituent of the sample. With an automatic FIA system a sample throughput of 200 to 300 samples per hour or even higher is easily attained.

In addition to the financial advantages, automatic analyzers offer advantages that may be termed as analytical. These are manifested in the better quality of the analytical results. As the samples and standards are treated in the system in a reproducible way and calibration may be checked easily by interspersing standards among samples, the results have a better reproducibility and less affected by personal errors than those yielded by the analogous manual method. This is true for all automated techniques. In addition to this, in the case of continuous flow methods, a further beneficial effect is observed, namely that most detectors, whether potentiometric, amperometric, or photometric, provide more reproducible results when used in flowing solutions compared to the batch mode. This results in a higher signal/noise ratio and a consequent improvement in detection limit (by a factor of 5 to 10).

A further advantage is the gain in analysis time. Due to the highly reproducible treatment of samples and standards, the analytical reactions need not be taken to completion to ensure the required precision of the results. The property of the flow technique that volumes, reaction times, and mixing patterns are controlled allows procedures that cannot be performed manually with the required precision either because the reactions do not reach equilibrium within reasonable time or because the products of reaction are unstable, to be carried out using flow analytical technique.

The advantages of flow analyzers, both financial and analytical, can be utilized using a FIA system even if manual injection is applied; therefore, the technique can be used in small laboratories with restricted financial means.

23.1. Analysis with Segmented Flow-Type Analyzers

23.1.1. Determination of Phosphorus in Plant Material

Principle of the Determination _____

Any technique that quantitatively converts the phosphorus content of the plant material can be used for the digestion. Phosphate ions form, in strongly mineral acidic medium in the presence of an excess of molybdate ions and of a given concentration of vanadate ions, a yellow-colored heteropoly acid:

$$H_{3+x}\left[PMo_{12-x} V_x O_{40}\right] \qquad (x = 1 \quad or\ 2)$$

In a medium 0.5 to 1.5 normal in acid the absorbance of the complex, as measured at 400 to 440 nm, is a linear function of the phosphate concentration in the solution.

Apparatus

Contiflo apparatus with photometer unit
Filter paper
Glass funnels
Erlenmeyer flasks, 500 cm^3
Erlenmeyer flask, 1000 cm^3

Chemicals

Sodium lauryl sulfate, purissimum
Ammonium meta vanadate, purissimum
Ammonium molybdate, purissimum
Disodium hydrogen phosphate, analytical grade
Perchloric acid, purissimum

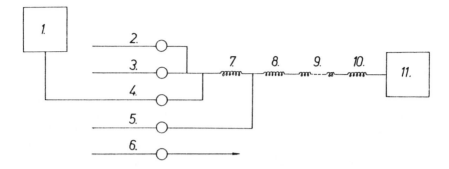

FIGURE 23.1

Flow diagram for the determination of the phosphorus content of plant material. 1: Sample; 40 samples/hour or 20 samples/hour, 2: air, 0.32 cm³/min, 3: acidic vanadate solution, 1.00 cm³/min, 4: sample, 0.10 cm³/min, 5: molybdate solution, 0.42 cm³/min, 6: to waste (from colorimeter), 0.80 cm³/min, 7: 1 cm³ mixing spiral, 8: 1 cm³ mixing spiral, 9: 4.5 cm³ delay spiral, 10: 1 cm³ mixing spiral, 11: colorimeter, 400 nm (400 to 440 nm).

Necessary Solutions

Diluting and rinsing solution: water

Acidic vanadate solution: 0.6 g ammonium vanadate and 100 cm³ concentrated perchloric acid/dm³

Molybdate solution: 15 g/dm³ ammonium molybdate

Sodium lauryl sulfate dissolved in these solutions so as to result in a concentration of 1 g/dm³

Standard Solutions

Solutions containing disodium hydrogen phosphate corresponding to 0, 2, 5, 8, and 10 ppm of phosphorus as well as acid and other additives are used for digestion of the plant material at the same concentration as in the samples.

Procedure

The flow system is assembled according to Figure 23.1 and the measurement is carried out in accordance with the general procedure described in the foregoing.

Evaluation

The unknown concentration is calculated as ppm phosphate in the solution further in % phosphate in the plant material (in the latter case, consider that 0.500 g dry plant material was digested and finally diluted to 100 cm³).

Calculate the scattering of the mean value. On the basis of all sets of data obtained, calculate the scattering of the technique in relative percent units.

23.1.2. Determination of Nitrogen in High-Protein Fodders

Principle of the Determination _____

The solution obtained in the Kjeldahl digestion of fodder, which contains the nitrogen in the form of ammonium salt, is made to react in an alkaline medium with sodium salicylate and sodium hypochlorite in the automatic analyzer system. Production of the indophenol-type dye, whose maximum absorbance occurs at 650 nm, is promoted by the presence of cupric sulfate catalyst and by heat treatment at 90°C.

The mechanism of the production of this compound is not totally clear yet. A number of oxidation and chlorination reactions are going on in parallel, and their relative rates are highly dependent on the conditions of the reaction; accordingly, this sensitive reaction can be utilized only in a mechanized system for the measurement of ammonium ions.

Apparatus

Contiflo apparatus with photometer unit
Graduated cylinder, 100 cm^3
Filter paper
Glass funnels
Erlenmeyer flasks, 500 cm^3
Erlenmeyer flask, 1000 cm^3

Chemicals

BRIJ-35 (polyoxyethylene lauryl ether)
Ammonium sulfate, analytical grade
Sodium salicylate, purissimum
Sodium hydroxide, purissimum
Potassium sodium tartrate, purissimum
Cupric sulfate, analytical grade
Sodium hypochlorite (commercial grade)

Necessary Solutions

Rinsing solution: water

Salicylate reagent: 0.5 M sodium salicylate solution

Alkaline tartrate solution: 0.05 M potassium sodium tartrate in 0.5 M sodium hydroxide

Sodium hypochlorite solution: a solution of 16 to 18 g/dm^3 active chlorine content

0.03% BRIJ added to all of these solutions (0.1 cm^3 30% BRIJ solution to 100 cm^3 solution)

Standard Solutions

Solutions containing ammonium sulfate corresponding to 0, 50, 100, 200, and 400 ppm N as well as acid and other additives are used for digestion of the plant material at the same concentration as in the samples. (In the calculation consider the mean loss encountered in the procedure.)

Procedure

Assemble the flow system in accordance with Figure 23.2 and carry out the measurement in accordance with the general procedure described in the foregoing.

Evaluation

The concentrations of the unknown solutions are calculated as ppm N in the solution and percent protein in the material weighed in (considering that 0.500 g dry plant material was digested and the solution was finally diluted to 100 cm^3). Calculate the scattering of the mean of the results. Compare the obtained values with those obtained from the same samples with the micro-Kjeldahl distillation technique and decide whether the results are significant at a 99.7 and a 95% reliability level.

23.1.3. Determination of Nitrite with Biamperometric Technique

Principle of the Determination _____

Nitrite ions are made to react with an excess of iodide ions in an acidic medium, whereupon an amount of iodine corresponding to the following equation is formed:

$$2\,NO_2^- + 2\,I^- + 4\,H^+ = 2\,NO + 2\,H_2O + I_2$$

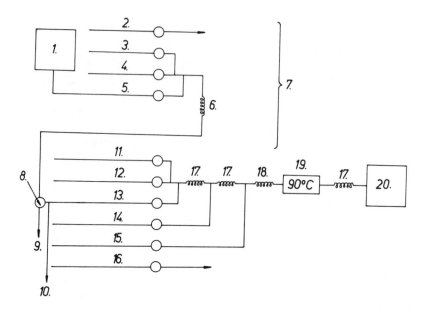

FIGURE 23.2

Flow diagram for the determination of the nitrogen content of plant material. 1: Sampler, 40 samples/hour, 2: water sampler (rinsing liquid) 2.00 cm^3/min, 3: air, 0.32 cm^3/min, 4: diluting solution, 0.80 cm^3/min, 5: sample, 0.10 cm^3/min, 6: 0.6 cm^3 mixing spiral, 7: dilution unit, 8: bubble separator, 9: to waste, 10: branching to P measurement, 11: air, 0.32 cm^3/min, 12: alkaline tartrate solution, 1.00 cm^3/min, 13: diluted sample, 0.23 cm^3/min, 14: salicylate reagent, 0.80 cm^3/min, 15: sodium hypochlorite solution, 0.23 cm^3/min, 16: to waste (from colorimeter), 0.80 cm^3/min, 17: 1 cm^3 mixing spiral, 18: 3 cm^3 delay spiral, 19: thermostat, 20: colorimeter, 640 nm (620 to 670 nm).

However, nitrogen monoxide produced in the reaction acts as a catalyst in the oxidation of iodide ions by atmospheric oxygen:

$$O + 2\,I^- + 2\,H^+ = I_2 + H_2O$$

Parameters influencing the reaction rate can be kept at a constant and identical level in case of the standards and the samples in the samples in the Contiflo analyzer. Therefore, the determination of the nitrite concentration is made possible even in the presence of air on the basis of the measurement of the amount of total iodine produced in the two reactions. In this manner, the sensitivity of the determination can considerably be increased.

The voltammetric (biamperometric) technique is used for the measurement of iodine. A DC voltage is applied across the platinum electrodes of the simple flowthrough cell inserted into the flow system and the current flowing through the cell is continuously measured with a potentiometric recorder.

By inserting a reductor into the flow system, automatically controlled $NO_3 \rightarrow NO_2$ reduction can be carried out and consequently the measurement of nitrate concentration can also be carried out in this manner. In practice, in the examination of plant materials and fertilizers, this task is the one that occurs most frequently.

Apparatus

Contiflo apparatus with voltammetric supply unit and detector
Erlenmeyer flasks, 500 cm^3
Erlenmeyer flask, 1000 cm^3

Chemicals

Potassium hydrogen iodate, analytical grade
Potassium nitrate, analytical grade
Potassium iodide, purissimum
Hydrochloric acid, concentrated, purissimum
Ammonium chloride, purissimum
Ammonium hydroxide, concentrated, purissimum
BRIJ 35 (polyoxyethylene lauryl ether)

Necessary Solutions

Rinsing solution: water
Potassium iodide solution: potassium iodide dissolved in
  ammonium chloride-ammonium hydroxide buffer of
  pH 9, 6, 100 g/dm^3
Buffer solution: 50 g ammonium chloride and 150 cm^3
  ammonium hydroxide/dm^3
Hydrochloric acid solution: 1.5 M
0.03% BRIJ added to the rinsing and the potassium iodide
  solutions.

Standard Solutions

0, 1, 2, 5, and 8 × 10^{-5} N sodium nitrite solutions dissolved in buffer solution diluted with water to 1:1; potassium hydrogen iodate solutions of the same (M/12) normality.

Procedure

The flow system is assembled according to Figure 23.3 and the measurement is carried out in accordance with the general procedure described in the foregoing.

Evaluation

Plot the iodate and nitrite concentration vs. current curves.

Calculate the concentrations of the unknown nitrite solutions at 99.7 and 95% statistical confidence limits.

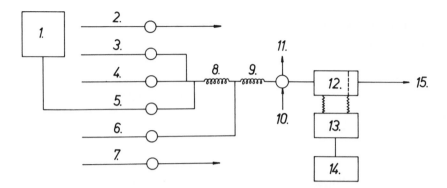

FIGURE 23.3

Flow diagram for the determination of nitrite. 1: Sampler, 20 samples/hour, 2: water sampler (rinsing liquid) 2.00 cm³/min, 3: air, 0.32 cm³/min, 4: potassium iodide solution, 0.32 cm³/min, 5: sample, 0.32 cm³/min, 6: hydrochloric acid solution, 0.80 cm³/min, 7: to waste (from biamperometric cell), 1.00 cm³/min, 8: mixing spiral, 9: mixing spiral, 10: bubble separator, 11: to waste, 12: biamperometric cell, 13: voltammetric detector unit, 14: potentiometric recorder, 15: to pump (connected to point 7).

23.1.4. Determination of the Potassium and Sodium Content in Soil Samples

Principle of the Determination

The determination of sodium and potassium is carried out with a two-channel flame photometer. Lithium chloride is applied as an internal standard. A soil extract prepared with ammonium lactate is measured. Wavelength: Na: 324.7 nm; K: 213.8 nm.

Apparatus

Contiflo analyzer with two-channel flame photometer
Volumetric flask, 1000 cm³
Erlenmeyer flasks, 500 cm³
Erlenmeyer, flask, 1000 cm³
Graduated cylinder, 1000 cm³

Chemicals

BRIJ-35 (polyoxyethylene lauryl ether)
Lithium chloride, analytical grade
Lactic acid, hydrolyzed, analytical grade
Ammonium acetate, analytical grade
Acetic acid, 96%
Sodium chloride, analytical grade

Necessary Solutions

> 30% BRIJ solution: (applied with *all* reagents at an amount of 0.3 cm³/dm³)
>
> Ammonium lactate solution: 32 cm³ hydrolyzed lactic acid, 7.7 g ammonium acetate, and 18 cm³ 96% acetic acid dissolved in distilled water; the solution is made up to 1 dm³
>
> Rinsing solution: ammonium lactate solution (see previous section)
>
> Diluting solution I, II: lithium chloride solution, 0.2%

Standard Solutions

1. Sodium and potassium stock solution: 0.53 g KCl and 2.542 g NaCl dissolved in water and the solution is made up to 1 dm³.

2. The stock solution (1) is diluted with ammonium lactate solution so as to obtain solutions corresponding to 5, 10, 15, 20, and 25 ppm concentrations.

Procedure

Assemble the flow system in accordance with Figure 23.4 and carry out the measurement in accordance with the general procedure described in the foregoing.

Evaluation

Calculate the sodium and potassium content of the unknown solution in ppm and as percent of the sample taken. Also calculate the scattering of the mean.

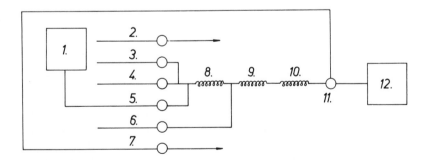

FIGURE 23.4

Flow diagram for the determination of potassium and sodium content in soil. 1: Sampler, 40 samples/hour, 2: water to rinse sampler, 2.00 cm³/min, 3: air, 0.32 cm³/min, 4: water, 1.40 cm³/min, 5: sample, 2.90 cm³/min, 6: water, 1.40 cm³/min, 7: suction from bubble separator, 1.00 cm³/min, 8: mixing spiral (2 × 10 windings, 18 mm diameter), 9: mixing spiral (3 × 10 windings, 18 mm diameter), 10: mixing spiral (5 windings, 18 mm diameter), 11: bubble separator, 12: flame photometer.

23.2. Analysis with FIA Systems

23.2.1. Determination of Phosphorus in Plant Material Following Dry Ashing or Digestion by Flow-Injection Spectrophotometry

Principle of the Determination

The phosphorus content of the plant material is converted to orthophosphate by ashing using $Mg(NO_3)_2$ as ashing aid or by digestion with HNO_3 in a closed vessel. In the reaction of orthophosphate with ammonium molybdate in the presence of strong mineral acid yellow-colored heteropoly-acids are formed of which $H_3P(Mo_3O_{10})_4$ is the best known species. Heteropoly-acids can be reduced to molybdenum blue in which part of the molybdenum has +5, and part +6 oxidation number. The composition of the heteropoly-acid as well as that of the reduced compound depends on the reaction conditions (pH, concentration ratios, etc.). However, under well-defined conditions the concentration of orthophosphate can be determined based on the absorbance of the reduced product measured at 670 nm.

Sample Preparation

1. Digestion with concentrated nitric acid in a closed vessel at 150°C.

Apparatus

Teflon-lined digestion vessel (pressure bomb) heating block or oven

Chemicals

cc HNO3

Procedure

Weigh about 0.1 g of the air-dried sample on an analytical balance and transfer it to the digestion vessel. In a fume cupboard add 1 ml of cc HNO_3, mix with the sample, and allow to stand for about 10 min. Close the vessel and in a heating block or oven heat at 150°C for 2 h. After cooling to room temperature open the vessel in a fume cupboard using vinyl gloves and leave the vessel there until the red nitric oxide is removed. Transfer the residue to a 100-ml volumetric flask with small portions of distilled water and dilute to the mark. If necessary, filter the solution to remove any solid particles.

2. Ashing with $Mg(NO_3)_2$ ashing aid at 500 to 510°C.

Apparatus

Porcelain crucible
Porcelain triangle
Muffle furnace

Solutions

$Mg(NO_3)_2$, 50% (w/w)
HCl, 2 mol/l

Procedure

Weigh about 0.1 g of the powdered air-dried sample into a porcelain crucible
on an analytical balance, add 1.00 ml of $Mg(NO_3)_2$ solution, place crucible over
Bunsen burner on a porcelain triangle in a fume cupboard, and heat slowly to
remove water and nitric oxide until no change occurs on further heating. Place
the crucible into a muffle furnace and heat at 500 to 510°C for 2 h. Allow to
cool in a desiccator and dissolve the residue in 5.00 ml 2 mol/l HCl. Transfer
the solution to a 100-ml volumetric flask and dilute to the mark with distilled
water. If necessary, filter the solution to remove any solid particles.

Determination of Phosphate

Determine the orthophosphate content of the solutions prepared by photometry
using the flow-injection principle.

Apparatus

Flow-injection manifold with photometer and recorder
(Figure 23.5)
Suction flask or ultrasonic bath

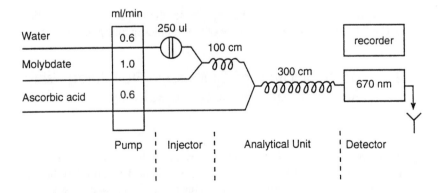

FIGURE 23.5
FIA manifold for the determination of phosphate.

Solutions

1. Molybdate solution
   3.5 g $(NH_4)_6Mo_7O_{24} \cdot 4\ H_2O$
   0.08 g $KSbOC_4H_4O_6 \cdot 1/2\ H_2O$ (potassium antimonyl tartrate acting as catalyst in the reduction to molybdenum blue)
   70 ml cc H_2SO_4 (dilute to 1000 ml with distilled water)

2. Reducing solution
   5.2 g ascorbic acid dissolved in distilled water and diluted to 200 ml

3. Phosphate stock solution (1000 mg/l PO_4-P)
   5.7464 g $Na_2HPO_4 \cdot 2H_2O$ dissolved in distilled water and diluted to 1000 ml

4. Series of solutions for calibration
   0.1, 0.2, 0.5, 1.0, 1.5, 2.0, 2.5, and 3.0 mg/l PO_4-P; prepare from the stock solution by dilution with distilled water

Measurement

Before use, deaerate about 300 ml of distilled water and 200 ml of molybdate solution by sonication or by suction with a water jet for 15 min. Pass deaerated distilled water through the flow injection system for about 10 min, then switch to the molybdate reagent and reducing solution and run the apparatus for 5 min; make the necessary instrument adjustments. Inject calibration solutions and samples in duplicate.

Calibration Curve

Prepare the calibration curve by plotting the peak heights vs. concentration of the calibration solutions.

Evaluation

Determine the concentration of the sample solutions using the calibration curve. Calculate the phosphorus content of the samples in milligrams per gram.

23.2.2. Determination of Glucose in Fruit Juice Using Amperometric Glucose Electrode

Theory

The flow-injection technique has two important features which make it extremely advantageous when biosensors are applied for detection. On the one hand in FIA methods a relatively short contact time is allowed between the sample and the detector. In the time between subsequent injections a background solution optimal for sensor stability is streamed through the analytical channel. On the other hand, FIA technique provides a fast and easy way for system recalibration or sensor stability check.

Bioelectroanalytical sensors are used in cells of different geometries. In the present experiment a cylindrical, flat-bottomed universal cell is connected to the end of a conventional single-channel FIA manifold. The solution enters the cell in the center of the bottom. The measuring surface of the biosensor (enzyme electrode) is centered close above the inflow orifice, leaving only a thin gap between the cell bottom and the sensor. The solution flows radially along the electrode surface and then further up to the broader part of the cell where the reference and the counter electrode are situated. As the top of the cell is open, it is easy to introduce and position the electrodes, and air bubbles have an easy way out. A suction tube carries the excess solution to waste. The position of the tube determines the solution level in the cell.

The method for glucose determination is based on amperometric detection by an enzyme electrode. The enzyme electrode is basically a platinum disk covered with a reaction layer containing glucose oxidase enzyme. A potential difference of 0.7 V is applied between the enzyme electrode and an Ag/AgCl reference electrode. In a background electrolyte in the absence of glucose a small current, the residual current flows. In the presence of glucose the following reaction takes place in the reaction layer:

$$\beta\text{-D-glucose} + O_2 \xrightarrow{\text{glucose oxidase}} \text{gluconic acid} + H_2O_2$$

The base-sensing element, the platinum disk electrode, detects the local concentration of H_2O_2.

In the flow injection system small volumes of glucose-containing solutions are injected into a stream of background electrolyte, and peak-shaped current vs. time curves are recorded.

Apparatus

Potentiostat with current-time recording device
Measuring cell (as described above)
Glucose enzyme electrode (prepared as described in Chapter 7)
Ag/AgCl reference electrode
Platinum counter electrode
FIA manifold (Figure 23.6)

Solutions

Phosphate buffer (pH = 7.3)
Glucose stock solution, 10 mmol/l; prepare 1 day before the experiment to allow mutarotation to reach equilibrium
Glucose calibration solutions (0.1, 0.5, 1.0, 3.0, 4.0, and 5.0 mmol/l); prepare from the stock solution by dilution with water

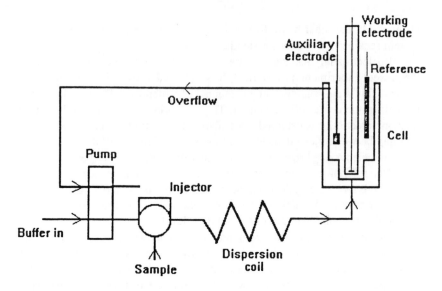

FIGURE 23.6
FIA manifold for the determination of glucose.

Procedure

Put together the FIA manifold, attach the measuring cell, and connect the electrodes to the potentiostat. Start pumping the background solution and record the current. After a sharp increase the current gradually decreases and after 10 to 15 min a steady current is established. Then inject the standards and samples in duplicate. Dilute samples whose glucose concentration falls outside the calibration range and repeat the injections.

After use, remove the enzyme electrode from the cell and store in buffer solution in the refrigerator.

Evaluation

Prepare the calibration graph by plotting the peak height against the glucose concentration. Determine the glucose concentration of the samples using the calibration graph.

23.2.3. Determination of Acid and Fluoride Content in Rain Water

Principle of the Ion-Selective Electrode-Based FIA Method _____

Flowthrough analysis has opened a new way of automation of the analysis of a series of samples of similar composition. In devices based on the flow-through analysis principle the sequences of the analysis are performed in succession in a flowthrough channel. For detection flowthrough detectors — mostly in continuous operation mode — are used.

The compatibility of the flowthrough principle and the properties of the potentiometric ion-selective electrodes is an important question. Two problems must be studied in this respect: the behavior of the detectors under flow-through conditions on the one hand, and the conditions required for incorporating such sensors in the flow-system on the other hand.

With respect to the behavior, electroanalytical detectors are especially suited for use in flow-through conditions. An important advantage is, for example, that the chemical and mechanical interferences, i.e., washing, cleaning, changing the solutions, etc., are significantly smaller in flow-system as compared to classical stationary measuring conditions.

In flowing media the compounds formed at the electrode-solution interface and materials dissolved from the electrodes are physically removed from the solution volume around the electrodes and thus have no disturbing effect. Another advantage is that, contrary to classical measuring conditions, convection in flow systems advantageously affects the response time of the potentiometric electrodes. Under flow conditions the detection limit of the electrochemical sensors in general is improved.

Lastly, in flow analytical methods the reference electrode can be placed downstream, thus eliminating the contaminating effect of the salt bridge electrolyte and ensuring a stable continuously renewing sample-reference electrolyte interface. The latter is especially important in potentiometric measurements.

The drawback of potentiometric measurements in flow methods can be the development of an oscillating analytical signal due to the pulsation of the solution carrying peristaltic pump and of the streaming potential arising in solutions of low ionic strength. The streaming potential can be calculated with the Helmholz-Smoluchovsky equation as:

$$E = 3.2 \frac{v\phi\varphi l}{\pi^2 d^2 \lambda}$$

where v is volumetric flow rate, φ is dielectric constant, ϕ is the electrokinetic potential, l is the length of the solution carrying tube, d is diameter of the tube, and λ is the conductivity of the electrolyte.

Acidity and Fluoride Concentration Determination in Rain Water

It has been shown for acid precipitation (pH 4.5) by a great number of measurements that the acidity is in most cases caused by strong mineral acids (H_2SO_4, HNO_3). Therefore, other components present in low concentration have only a slight effect on the buffer capacity and ionic strength of the sample, and do not significantly influence the acidity itself. Accordingly, with such samples, dilute aqueous solutions of strong mineral acids can be used as standards; the pH of both the standards and samples are measured with a pH glass electrode and the acid concentration of the sample is expressed.

The determination of fluoride in rain water by ion chromatography poses a problem due to the presence of formate and acetate. The flow-injection technique based on a the highly selective lanthanum fluoride single crystal electrode is an obvious solution of the problem. However, the flow-injection technique should have a detection limit similar to that reported for fluoride by ion chromatography.

At selecting the flow-injection conditions for the determination of hydrogen and fluoride ions, the unfavorable dynamic characteristics (slow response) of the relevant ion-selective electrodes at low concentrations or in solutions of low buffer capacity were considered. By increasing the volume of solution injected and decreasing the volumetric flow rate in the system the residence time of analyte in the detector cell is increased. Moreover, the introduction of analyte ions into the carrier stream in a relatively low concentration has the same role in addition to ensuring a stable baseline.

The rate of the analysis of ionic constituents in wet precipitation can be increased by applying a nonequilibrium "dynamic potentiometric method," e.g., the flow-injection technique. When the flow-injection technique is used, the potentiometric signal measured depends primarily on the pH and fluoride concentration of the solution, but it is affected also by the hydrodynamic parameters of the flow system and the dynamic characteristics of the detector. Thus, analytical data can only be obtained under controlled experimental conditions.

Apparatus

A peristaltic pump with a variable flow rate; a six-port manual injector provided with loops of different volumes; a 20-μl loop for hydrogen injection and a 250-μl loop for fluoride determination

Dispersion tube of 1-m length of Pierce Teflon$^{\circledR}$ tube with internal diameters of 0.3 and 0.6 mm for acid and fluoride determination, respectively

A pH-sensitive, flowthrough microcapillary glass electrode and a flowthrough fluoride electrode used as an indicator electrode, double-junction Ag,AgCl electrode employed as reference electrode; salt bridge electrolyte is 0.5 M KNO_3

Potential measurements made with a digital pH-mV meter; signals recorded by an x-t recorder

Chemicals

Stock solutions prepared from analytical-grade hydrochloric acid and sodium fluoride with doubly distilled water; standard solutions made by serial dilution

Carrier solution: for the *acid concentration determination* 0.5 M KNO_3 solution containing a low concentration ($<10^{-4}$ M, e.g. 5×10^{-5} M HAc/NaAc buffer solution) of buffer was used

Acid standards: 10^{-5}; 2.5×10^{-5}; 5×10^{-5}; 10^{-4}; 2×10^{-4}; 5×10^{-4}; 10^{-3} *M* HCl or Britton Robinson buffers diluted hundred times

For the *fluoride determination,* samples and standards diluted in 1:1 ratio with TISAB solution; the TISAB solution used contained 3.07 g l^{-1} EDTA, 9 g l^{-1} sodium chloride, and an acetate buffer with a total acetate concentration of 1.5 *M,* adjusted to 5.0 to 5.3

The *carrier solution* prepared by diluting the above TISAB solution with equal volume of water and adding sodium fluoride solution to yield the required low concentration (10^{-6} *M*) of fluoride

Fluoride standards: 10^{-7}; 2×10^{-7}; 3×10^{-7}; 5×10^{-7}; 6×10^{-7}; 8×10^{-7}; 10^{-6} *M* NaF in TISAB

Procedure

Set up the flow injection manifold as shown in Figure 23.7.

The carrier solution is streamed with a volumetric flow rate of 1.3 ml/min, while the salt bridge electrolyte with that of 0.3 ml/min.

Estimate the concentration sensitivity and the detection limit of the electrodes on the basis of the steady-state signals obtained by pumping through the system the appropriate calibration standards with a volumetric flow rate of 1.3 ml/min.

Flow injection calibration is carried out after obtaining a stable baseline in the appropriate carrier solutions at injecting successively the relevant calibration standards, each of them three times. Plot the calibration graph; $\Delta E_{max,average}$, vs. log $c_{injected}$. Inject each sample at least three times and make the average of ΔE_{max} measured. Check the baseline shift by the alternative injection of standards after the analysis of five samples.

Use the direct potentiometric "calibration graph method" for *evaluation* of the ΔE_{max} measured for the samples.

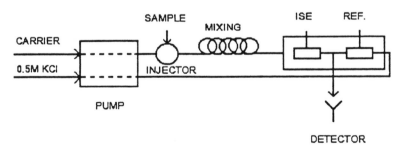

FIGURE 23.7
Manifold for flow-injection potentiometric measurement with hydrogen and fluoride ion-selective electrode, respectively.

23.3.   References

Proc. Technicon International Congress, Mediad Inc., White Plains, NY, 1970.

Karlberg, B. and Pacey, G. E., *Flow-Injection Analysis (A Practical Guide),* Elsevier, Amsterdam, 1989.

Ruzicka, J. and Hansen, E. H., *Flow-Injection Analysis,* Wiley-Interscience, New York, 1981.·

Valcárcel, M. and Luque de Castro, M. D., *Automatic Methods of Analysis,* Elsevier, Amsterdam, 1988.

Index

A

Acetic acid, determination
  by conductometry, 48
  by oscillometry, 53
  of dissociation constant, 7
Acetyl groups, determination
  by ion exchange, 209
Acid, determination
  by flow-injection analysis (FIA),
    374
Activation analysis, 340
  neutron (NAA), 340, 347
Activity coefficient, 10
Alcohols, determination
  by gas chromatography (GC), 230
Aldehyde, identification
  by thin-layer chromatography
    (TLC), 217
Alkoxy groups, determination, 198
Alkyl benzene homologs, identifi-
    cation
  by GC, 228
Amino acids, separation
  by high performance liquid
    chromatography (HPLC), 308
p-Amino-azobenzene, determina-
    tion
  by TLC, 220
Amperometry, 60
Ampoules, examination of

by oscillometry, 55
Anions, determination
  by ion chromatography, 277–279
Antibiotic materials, determination
  by HPLC, 303
Antioxidants, determination
  by HPLC, 301
Atomic absorption spectroscopy
  cold vapor, 125, 132
  flame, 117, 127
  graphite furnace, 119, 129
  principles, 115
Automatic analyzers
  principle, 359
Azo-compounds, identification
  by TLC, 217

B

Bauxite, analysis
  pyrite in, 169
Beer's law, 73
Biamperometric titration
  determination of iodine, 17
  principle, 17
Biocatalytic sensors, 59
Blood serum, analysis
  drugs in, 293, 299
  iron in, 89
  potassium in, 10
Blood plasma, analysis, 299